BIOMEDICAL PHYSIOLOGY

Revised Printing

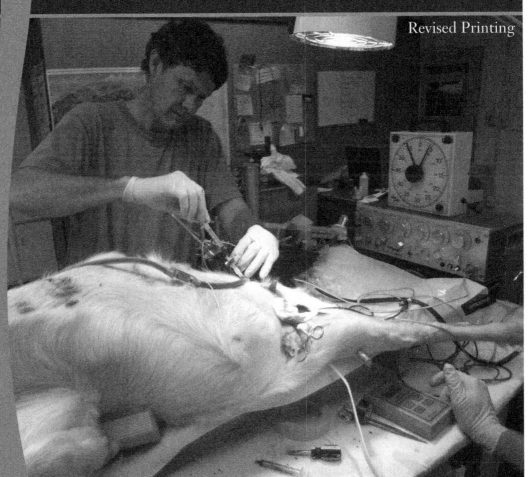

MATTHEW L. GOODWIN ● LAWRENCE C. WIT

Weill Cornell Medical College   *Auburn University*

Kendall Hunt

publishing company

Figures 15.2 and 15.10 are illustrations by Jamey Garbett.
© 2003 Mark Nielsen

All other interior images © Kendall Hunt Publishing Company

Cover image © courtesy of the author.

Kendall Hunt
publishing company
www.kendallhunt.com
Send all inquiries to:
4050 Westmark Drive
Dubuque, IA 52004-1840

Copyright © 2011 by Matthew L. Goodwin and Lawrence C. Wit
Revised printing 2012
ISBN 978-1-4652-0933-7

Printed in the United States of America
10 9 8 7 6 5 4

CONTENTS

For videos from Dr. Goodwin explaining some of the concepts presented in this book (and other concepts), please visit:

http://www.youtube.com/user/DRphysiologyguy

ACKNOWLEDGMENTS

Now in a newly revised printing, this text is a unique culmination of many things. First, we must thank Kendall Hunt for publishing an unusual work. There are many very good (exhaustive and review) physiology books on the market, but not a text like this (see Introduction). Kendall Hunt recognized the uniqueness and importance of this text and was extremely accommodating in working with both of our busy schedules as an Associate Dean & Professor of the College of Sciences and Mathematics at Auburn University (LCW) and a medical student at Weill Cornell Medical College (MLG). We cannot thank Kendall Hunt's Stefani DeMoss and Katie Wendler enough for their patience and guidance throughout this project. A special thank you goes to Amanda Smith of Kendall Hunt for her work on this revised printing.

Second, we cannot take too much credit for the work here. Although the material is directly from (and used for) Larry Wit's well-known Mammalian Physiology Course at Auburn University and was formally written into this text by Matthew Goodwin, many previous versions of some of the notes included here were already circulating amongst students. So we thank the previous students who contributed to those notes, and thus to the development of this text. Although it is hard to tease out exact contributions to the various notes that have circulated, we wish to specifically thank former Kinesiology graduate students and former Pharmacology graduate students, although many others contributed. A special thank you goes to Dr. Ginny Fuller, as she made many early corrections to the manuscript and provided helpful feedback. We also must thank those who volunteered to edit individual chapters for both grammar and content: Dr. Kevin A. Crosby, Adam S. Faye, Vinay L. Patel, Dr. Ankit B. Patel, Neha Kumar, Dr. Sandeep (Sunny) P. Kishore, and Anthony J. Choi. Also, a special thank you to Dr. Wit's Spring 2012 students, who edited the first edition of this text, and in particular Rucker Staggers.

A project of this scope is not possible without many great teachers of physiology throughout our careers.

First, I (MLG) must thank Mandy, my partner of many years, for her love and support throughout this and all the other projects concerning science, medicine, and education that are so important to me. Many "vacations" have been spent working on projects that I am passionate about, like this one. This would not have been possible without her understanding, support, and sense of humor through it all.

I must also thank my my father, for encouraging me and providing my earliest "physiology reading material" as my interest in physiology began while I was a young high school runner, although we didn't know what shape that interest would take at the time. I also thank my mother, who taught me how to teach, and instilled in me a love of breaking complicated topics into comprehensible concepts that others could more easily learn. She also has been my role model as a true teacher my whole life - teaching to the students before her, always meeting them where they were in their knowledge and in their lives. I strive to be half the teacher she is. Although I caused him much angst, my high school cross country coach, Rick Zeller (Wade Hampton, Greenville, SC), should be given credit for (like many authority figures in my life) putting up with my constant questioning of "why" we were doing any particular workout, demanding evidence. Coach Zeller truly is one of, if not *the* best in the country at what he does. I must also thank the faculty of the Furman University Health Science department (and specifically Drs. Ray Moss, Tony Caterisano, and Bill Pierce) for fueling my interest in physiology and sparking my interest in lactate metabolism. I must also thank my friend Coach Chris Fox, who appropriately helped me channel my interest in physiology into research and medicine. He applies physiology and psychology as well as anyone I know.

I also wish to express my deep gratitude to co-author Dr. Larry Wit for both the instruction I received as a student of his in the classroom and for serving as a member on my PhD committee. If you were lucky enough to be in his class at some point, you know that his gift in the classroom is truly a rare one. Working with him on this project has been a true honor.

Finally, I want to thank my mentor, Dr. L. Bruce Gladden (cover), for his example as a scientist, teacher, mentor, citizen, man, and friend. More was gleaned from spending time in experimentation and discussion with Dr. Gladden than could ever be learned in a classroom. If you've been in his lab before, you know he is a special scientist, articulate in his expertise area as well as in a wide range of physiological concepts. I fear fewer and fewer scientists like him are being trained.

Last, I thank you if you were one of the many who engaged me in debate or discussion, and especially if you were one of my students who ever questioned me as a teacher and made me explain "why" by providing evidence; hopefully there are many more "problem" students like you out there.

Matthew L. Goodwin, PhD

For more than 35 years, I (LCW) have had the privilege of teaching physiology courses to college students. My training in physiology began in the early 1960s when I took an anatomy and physiology course in high school and was immediately taken with the simplicity and complexity of how the human body works.

I have continued to learn via my formal education in college and graduate school, but I have learned most as I have taught thousands of students through the years. My desire has always been to make the course more about learning to think physiologically rather than just memorizing lots of details. In the end, I hope that has been the case for my students; I know it has been the case for me. Sometimes I think I have learned far more than I have taught.

As Matt has said, we owe a debt of gratitude to those who have influenced us. In that regard, I would recognize Dr. Homer Dale, a physiology professor I had in graduate school. He is still the best teacher I have ever had. Also, I must mention Dr. John Pritchett, a physiologist and former colleague at Auburn. As a young professor, he influenced me greatly and was the progenitor of Mammalian Physiology, a course I now teach.

Finally, I would like to thank my wife, Nancy, for her understanding and unwavering support for nearly 40 years...they don't come any better!

Lawrence C. Wit, PhD

INTRODUCTION

"The profession in which human physiology is the most useful and for which it is now in fact the basic science is the medical."
-    August Krogh, *Science*, 1939, 89(2320); 545-7

Mentioned by August Krogh in 1939, it still holds true today: the foundation of medicine, the basic science upon which all of medicine rests, is human physiology. If a physician or other clinician does or does not know the name of some transcription factor or second messenger, he or she can look it up or still understand the whole-body response. However, if that same clinician does not understand human physiology, there is little chance that he or she can interpret new data about that patient correctly. Without a firm foundation in physiology of the human body, practicing any branch of medicine becomes impossible.

It is our hope that the pages that follow will assist in your forming a foundation of physiological knowledge that will last a lifetime. As you learn this material, constantly review and organize the information so that you can always "carry it around in your pocket with you." There is a lot of information here, and a semester learning this book will be difficult. However, if you intend to remain in medicine, what you will get out of really learning this material will be priceless.

In this book you will find that some charts, graphs, pictures, and questions have been purposefully left out or blank for you to complete during class discussions or on your own throughout the course of the semester. The only way to really learn mammalian physiology well is to use the material daily, as an aid to thinking critically and physiologically. The three things that are paramount to your successful mastery of this material are (1) knowing and understanding it all, backward and forward; (2) studying daily so that you never fall behind on the material; and (3) *participating in class questions and discussions on a daily basis.*

This book is unique in that it is not designed to be a primary text or a reviewing tool for any examination (such as review books for medical students). Rather, this book is designed specifically to provide a framework from which a student can understand biomedical physiology in one semester. When I (MLG) was a student in Dr. Wit's course, I began recording lectures and transcribing them at night to be sure I didn't miss anything. In addition, there was a desire for us as students to be more involved during class. Eventually, this turned into my taking the existing notes that were being passed around, correcting and rewriting them, and combining them with my notes and my knowledge of physiology—and this was the result. The impetus to publish this came from the notes receiving wide use both at Auburn and at other places (I was referring to them and distributing them often at Weill Cornell Medical College, for example). This was originally designed as a set of "notes" for Mammalian Physiology at Auburn University, and we've tried to keep it as true to this form as possible. To that end, this work is designed to be used during class and to be taught from. The margins should allow you to conveniently write in this, as is the intention. It also is our intention that you keep this text and refer to it and to the notes you make in the margins and other spaces as you move forward in your biomedical career.

We have also added some things that we feel make the text more useful:

1. Quizzes at the end of most chapters. These can be used however the course instructor wishes. These provide feedback as to the types of questions you should be able to answer at the completion of each section.

2. Suggested readings for each chapter. Although these are listed at the end of each chapter, it may be useful to read some of these before the sections begin. Other times the readings are simply there to reference if you desire. As you will see, we have found Guyton and Hall's *Textbook of Medical Physiology* to be a nice complementary, more exhaustive text. To reiterate, our book is designed so that you can learn it in one semester and have the basics of physiology "locked in" for a lifetime. To that end, we have purposely tried to maintain the "notes" format and not go into too much detail on any one topic.

If you find any errors or wish to comment on any of the material, please feel free to e-mail MATTHEWLGOODWIN@gmail.com. In this way we can ensure that future editions of the text continue to improve.

Finally, a word on how important physiology discoveries are made. In addition to its numerous critical uses in medicine, animal experimentation represents a critical and required method by which physiological discoveries are made. Although care is taken to reduce suffering and exhaust other means of discovery, examining physiology in an intact animal remains a required and critical step in both discovery and safety testing of novel therapies. Although computer models continue to improve and *in vitro* work directs early research efforts, animal experimentation provides the critical translational link by which discoveries can ultimately be used on humans. Unfortunately, fewer and fewer people seem to understand that their very health and existence rest on both past and current humane animal experimentation. With the recent explosion in genomic data available, the coming years will place a new premium on intermediate physiological models (like those employed by L. Bruce Gladden; cover photo) that bridge *in vitro* or murine work to human models.

For the sake of our children (and grandchildren), we hope we've trained enough translational physiologists to bridge this gap.

Matthew L. Goodwin, PhD
Lawrence C. Wit, PhD

CHAPTER 1

Body Fluid Compartments

I. **Water:** The human body is about 60% (~50% to 70%) water.
    a. The percentage depends on "fatness" or "leanness": in lean people a higher percentage of body weight is water, whereas in fat people a lower percentage of body weight is water.
    b. **Women** will have a slightly lower percentage of total weight as water because they have more essential fat.
    c. Two major water compartments in the body:
       i. **Intracellular**: Inside cells (~40% of weight)
       ii. **Extracellular**: Outside cells (~20% of weight)
          (1) **Interstitial fluid**: "Bathes" the cells (~15%)
          (2) **Plasma fluid**: Fluid component of blood (~5%). Remember, the fluid in blood is plasma—the formed elements are red blood cells (RBCs), white blood cells (WBCs), and platelets. **Hematocrit** is the percentage of blood volume that consists of the formed elements after spinning the blood in a centrifuge.
          (3) **Extracellular fluids that are minor categories** (neither plasma nor interstitial):
             a. **Lymph** fluid is returned from interstitial fluid back to plasma via lymphatic vessels, which make up a one-way circulatory system. This is how fats get absorbed from the digestive tract (into the lymphatic system). Those fats then get dumped into venous blood as lymph drains into the venous system.
             b. **Transcellular** fluid is secreted by special cells into some sort of cavity or space. An example would be synovial fluid at joints or cerebrospinal fluid (CSF).
       iii. **What about gastrointestinal (GI) fluids and urine?**
          (1) These are technically "outside" of the body! However, note that many physiologists might consider these fluids transcellular once they reach the stomach, where the makeup of the fluid is mainly determined by the secretion of HCl (pH ~2), *not* by what was eaten. However, for now, **we will not call GI fluids "transcellular."**

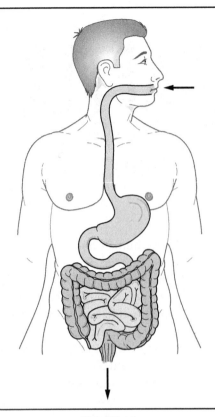

FIGURE 1.1

The composition of fluid compartments remains remarkably constant!

Claude Bernard, a French physiologist that some called the Father of Physiology, first recognized this "constancy of internal environment" in the 1800s.

Walter Cannon, an American physiologist, in 1932 coined the term "homeostasis" for this tendency of the human body toward a constant internal environment.

The great theme of physiology is homeostasis!!!

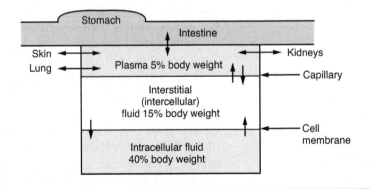

FIGURE 1.2

d. Boundaries between fluids
  i. **Capillary wall**: between plasma and interstitium
  ii. **Cell membrane**: between interstitium and cell cytoplasm
    (1) To change composition of any fluid compartment, you have to start at the plasma. The plasma is the only compartment in direct communication with the environment. Example: How does a hypoxic cell obtain O_2?
    (2) Fluid compartments remain relatively constant because of homeostasis and homeostatic mechanisms.

II. **Measurement of body fluids**
  a. **Dry weight determination**
    i. We could "cook" people after they die to determine the percentage of water. In many experimental labs, animal muscle is dehydrated in this manner to determine if there was any **edema** (excess fluid).
  b. **Dilution technique**
    i. We use this method today.
    ii. **For plasma volume**: Inject something that cannot cross the capillary membrane. Let it circulate, then draw a plasma sample and measure its concentration. Use an isotopically labeled plasma protein (e.g., albumin with label). Inject a known amount. After it circulates, the concentration will allow you to calculate the plasma volume. **If 100 mg labeled albumin is injected into an animal, and the concentration is determined to be 1 mg/mL, what is the plasma volume?**

    iii. **For interstitial volume**: Determine plasma volume first. Then inject something that can cross the capillary wall, but *not* the cell membrane (e.g., inulin). **We know plasma volume to be what we calculated above; we inject 300 mg of inulin and then measure a plasma concentration of 1 mg/mL inulin. What is the interstitial fluid volume?**

    iv. **For intracellular volume**: Determine above volumes. Then, inject something that crosses all barriers (labeled H_2O). **If we inject 700 mg of labeled H_2O and we know the above values (for plasma and interstitial fluid), and we then measure a plasma concentration to be 1 mg/mL, what is the intracellular fluid volume?**

III. **The Cell Membrane**
  a. **Structure:** Fluid Mosaic Model has dominated since 1970s. Lipids are fluid component, and proteins provide the mosaic quality.

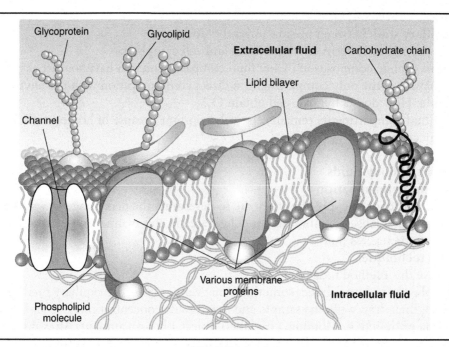

Glycoprotein   Glycolipid

Extracellular fluid   Carbohydrate chain

Lipid bilayer

Channel

Various membrane proteins

Intracellular fluid

Phospholipid molecule

FIGURE 1.3

b. The majority of the cell membrane is due to the presence of phospholipids.
  i. They are bipolar: one end of phospholipid is water soluble (hydrophilic—the phosphoric acid) and the other end is lipid soluble (hydrophobic—the lipid (FA) tail).
  ii. Hydrophilic ends point to the outside of the membrane and hydrophobic ends point toward each other to form a **bilayer,** as shown above.
  iii. Thus, **the membrane is permeable to lipophilic things** (CO_2, O_2, alcohol, FAs, steroid hormones). The membrane is not permeable to ions, glucose, etc.
  iv. Hydrophilic items sent via the circulation are fine until they get to the membrane, where they must find a way to transverse it. Conversely, hydrophobic items are more difficult to transport in the plasma but can permeate directly through the membrane once they arrive.
  v. The membrane is not static; it can undergo some movement.
    (1) Phospholipids can move around some.
    (2) Proteins can undergo some changes: tertiary and quaternary structures can change shape.
  vi. For example, **PIP-2** is a particular phospholipid we will discuss later.
c. **Cholesterol**
  i. Major constituent of plasma membranes.
  ii. Maintains proper fluidity in membrane.
  iii. Precursor for steroid hormone synthesis.
d. **Glycolipids**
  i. Lipids with CHO moiety attached. These can function as receptors and can be antigenic.
  ii. **Cholera (due to *Vibrio cholerae*):** a disease mainly in the developing world. Receptor for the cholera toxin is a glycolipid that is involved in transport of fluids in the GI tract.
  iii. **Blood groups:** antigens are glycolipids on the surface of the RBCs.

e. **Proteins**
  i. Proteins are embedded in the membrane. Some penetrate the entire width of the membrane. Two general types:
     (1) Integral (intrinsic)—tightly associated with membrane lipids. Cannot extract these without disrupting membrane. Most of these will span the entire width of the membrane, touching both the intracellular fluid and interstitial fluid.
     (2) Peripheral (extrinsic)—do not penetrate deeply, if they penetrate at all. Not bound tightly. Most are on intracellular side of membrane.
  ii. **Functions of proteins**
     (1) As channels or pores: Typically, integral proteins exist "shoulder to shoulder." This leaves a channel in the middle for lipid insoluble molecules to pass through. Pores (or channels) can open or close electrically (voltage-gated) or chemically (ligand-gated). A closed channel is simply when a protein member of that channel changes conformation in such a way as to close off the channel. Channels or pores are molecule specific. Also, size and charge are limitations. Ions or smaller may move through some pores, whereas glucose is too large!

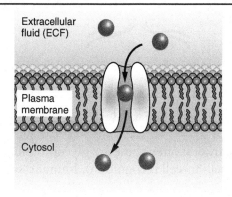

Extracellular fluid (ECF)

Plasma membrane

Cytosol

FIGURE 1.4

   (2) As transporters or carrier molecules: some proteins transport substances too big to move through pores. These are molecule specific. These proteins will change conformation/shape upon correct binding of molecule to enable that molecule to move across the membrane (much like a swivel chair). These can be saturated (Vmax) and may be poisoned (inhibited).

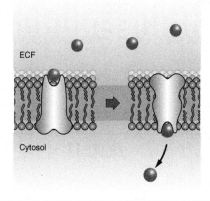

ECF

Cytosol

FIGURE 1.5

(3) **As receptors**: some proteins recognize substances to make the cell do something. Specific to a certain item (like a keyhole on a door).

(4) **As enzymes**: usually these proteins (catalysts) are located intracellularly.

(5) **As cytoskeleton**: these proteins provide the skeleton framework of cell or act as anchors to support the membrane.

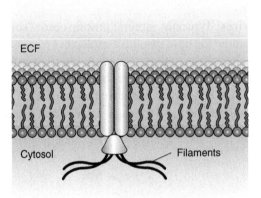

FIGURE 1.6

(6) **As cell identity markers**: often these are glycoproteins that communicate to an extracellular item (e.g., cell–to-cell recognition). Immune system recognizes various cells this way.

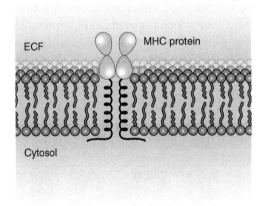

FIGURE 1.7

IV. Terms

a. **Glycocalyx**: fuzzy layer on cells due to glycoproteins and glycolipids.

b. **Adhesion molecules**: some integral proteins make contact with each other or other things in extracellular environment (cell to cell, or cell to matrix). Important for cell growth, cell differentiation, cell migration, and guiding cells. Dysfunction of adhesion is a hallmark of metastatic tumors.

c. **Autocrine secretion**: producer and recipient of cell signal are the same cell. May be used to turn off a cellular response in a negative feedback manner.

d. **Paracrine secretion**: different sender and receiver, but in close proximity. Neurons communicate with neurotransmitters in this way.

e. **Endocrine secretion**: different sender and receiver, but the target is a long way from sender. Usually uses the circulatory system.

f. **Signaling molecules** may be:
   i. Amines (epinephrine)
   ii. Peptides and proteins (cytokines and insulin)
   iii. Steroids (estrogen)
   iv. Small molecules (amino acids, nitric oxide)

V. How do cells communicate (transport information) across the cell membrane?

a. **Direct contact**: via gap junctions; involves connexons. (One gap junction between two cells involves two connexons. One connexon is made up of six proteins called connexins.) These are common in the heart, hepatocytes, smooth muscle, and many other places. These are *not* present in skeletal muscle. Communication involves cells very close to each other. Their protrusions touch each other, forming connexons. Information can be passed along (ions are passed, voltage is passed).

b. **Chemical signals**: different types
   i. **Ionotropic**: alter ionic permeability of channels and/or pores. We change conformation of channel door to open it. This can be very specific—voltage-gated or ligand-gated. This can be coupled with telling the cell to do something.
   ii. **G-Protein**: Seven transmembrane receptors (7TMRs) often are coupled with trimeric G-proteins that function as molecular switches to turn on some enzyme or other cascade event inside of the cell. Receptor binds substance on the outside and then signals a G-protein on the inside of the membrane to activate an enzyme, which may initiate a response/cascade.
      (1) **Cyclic nucleotides**: Adrenal medulla produces ~80% epinephrine; this epinephrine then goes to liver receptors (among other places). At the liver, it binds, activating an intracellular part of the G-protein ("Gs;" this alpha part of the G-protein is different, depending on the system). This activates adenyl cyclase (AC) to make cyclic adenosine monophosphate (cAMP). This allows protein kinases to be active, and eventually in a hepatocyte, glucose is released into circulation. This cAMP system has different results in different cells.
         a. Summary: Epinephrine (1st messenger) binds receptor → Receptor protein (integral protein) changes conformation → Gs (changes conformation) →Adenyl cyclase (turned on) → Converts ATP to cyclic AMP (2nd messenger) → Activates protein kinases → These phosphorylate specific intracellular enzymes → Conversion of glycogen to glucose…
         b. Phosphodiesterase catalyzes cAMP → 5'AMP. This shuts off the action of cAMP.
         c. Gs is found in the cyclic AMP systems; the receptor changes conformation to affect/stimulate the Gs so it can in turn affect the adenyl cyclase.

Postreceptor Event: Cyclic AMP Second Messenger System

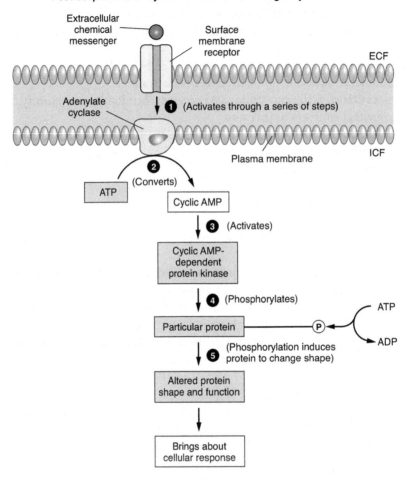

FIGURE 1.8

d. Epinephrine is too big to get into the actual cell. It is also lipid insoluble.

e. This system is a template for other systems and cells, for ways to turn on cell. **This system is an example of cyclic AMP signaling.**

(2) **Calcium Mechanism**: Hormone → Protein receptor (change conformation) → Gq (change conformation) → Intracellular enzyme activity (phospholipase C) increased → Affects phosphatidylinositol 4,5-bisphosphate (PIP2; a phospholipid in membrane) → Phospholipase C converts PIP2 into inositol trisphosphate (IP$_3$) and diacylglycerol (DAG) → IP3 migrates to endoplasmic reticulum (ER) and causes it to release Ca^{++} into cell → Ca^{++} does many things. In muscles → binds to troponin to allow muscles to contract, or in other tissues can bind to calmodulin to activate protein kinases. DAG may also activate some protein kinases.

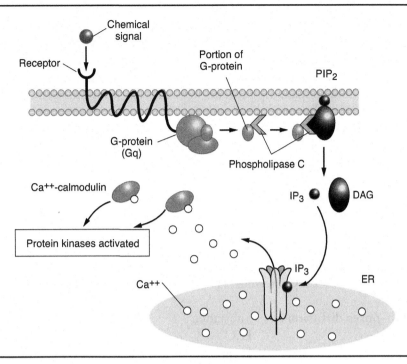

FIGURE 1.9

(3) Arachidonic Acid (AA) Mechanism: AA is typically found in membranes. G-protein here increases activity of phospholipase A2, which catalyzes the reaction shown in figure 1.10.

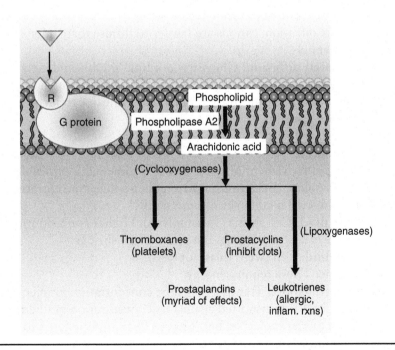

FIGURE 1.10

a. e.g., aspirin inhibits this conversion of AA to prostaglandins.
   iii. **Catalytic Receptors**: Receptor has three components: hormone-binding portion, membrane-spanning portion, and enzyme-containing portion. Tyrosine kinase mechanism: the mechanism by which insulin communicates with its target. Hormone alters conformation of receptor and activates tyrosine kinase, which phosphorylates tyrosine residues.
   iv. **Integrins**: Membrane-spanning proteins with a receptor on the outside leaflet and an anchor that goes through to the intracellular side.
   v. **Intracellular Receptors**: Lipid soluble; often steroids. Lipids can diffuse through the membrane, but they still need a receptor. This receptor is usually in nucleus. This then alters the genetic machinery.

VI. **How do we transport material across the membrane?**
   a. **Passive mechanisms:** No energy is required, but cannot transport against concentration gradient.
      i. **Simple diffusion:** Material movement from higher concentrations to lower concentrations through the membrane. Has to be lipid soluble and small.
         (1) Examples: O_2, CO_2, N_2, steroids, fat-soluble vitamins, urea, glycerol, small alcohols, ammonia
      ii. **Through-pores diffusion:**

Substance	Diameter A*	Diffusion
H_2O	3.0	5×10^7
Urea	3.6	4×10^7
Cl^-	3.86	3.6×10^7
K^+	3.96	100
Na^+	5.12	1
"Pore"	8.0	—
Glucose	8.6	0

*A= Angstroms

   1. As size increases, rate of diffusion decreases.
   2. Charges affect diffusion; cations are harder to diffuse than anions because Ca^{++} lines the outside of channels.
      iii. **Carrier-facilitated diffusion:** Glucose is not lipid soluble and is too big to go through pores. This is where something binds its carrier molecule, which then changes conformation, carrying it down its concentration gradient. **These carrier molecules**
         (1) are substrate specific,
         (2) can be saturated, and
         (3) can be inhibited.
      iv. **Osmosis:** The net movement of water through some semipermeable membrane. Water moves from low-solute concentration to a higher-solute concentration. If a cell is in:
         a. Isotonic fluid: Cell neither shrinks nor swells.
         b. Hypertonic fluid: Cell shrinks (plasmolysis). Higher concentration of solutes in the fluid outside of the cell, so water goes out and the cell shrinks.
         c. Hypotonic fluid: Cell swells (plasmoptysis).
   b. **Active mechanisms:** These require energy.
      i. **Primary ionic transport:** Transports against concentration gradient.
         a. Na/K pump: Na^+ is transported against its concentration gradient.
         b. These have a carrier molecule, but it has to be phosphorylated to be activated. It then changes conformation to transport.

ii. **Secondary active transport or Na$^+$ Co-transport:** Within the gut, nutrients (alanine (AL) and Na$^+$) are transported to epithelial cells by "front door" mechanisms that take advantage of Na$^+$ gradient. Na$^+$ goes down its gradient into epithelial cell. AL goes against the concentration gradient. Two transport sites on the transport molecule: one for AL (amino acid) and one for sodium. Na$^+$ is actively transported out the "back door" of the epithelial cell using ATP, so the Na$^+$ concentration in epithelial cell stays low, and there is always a concentration gradient to drive Na$^+$ at the front door. AL diffusion is driven by Na$^+$ passive movement down gradient into epithelial cell at front door and active transport out at back door.

(1) This is a combination of carrier-facilitated diffusion and direct ionic transport.

iii. **Bulk transport:** Transport in which large molecules or particles are transported. Two subparts:

(1) **Endocytosis:** Brings things into the cell.

a. **Phagocytosis:** The ingestion of large particles or microorganisms in specialized cells only. Requires a special stimulus, such as a large particle attaching to the cell.

b. **Pinocytosis:** Involves smaller particles and smaller vesicles than phagocytosis.

  i. **Fluid phase**: not very specific.

  ii. **Receptor mediated**: receptors in pit are specific.

(2) **Exocytosis:** The cell is putting something outside by packing it in a vesicle and extruding it outside the cell.

a. Neurotransmitters found in vesicles are released onto the surface of the cell. **This requires energy.**

VII. **Pathologies**

a. **G-proteins: alpha, beta, and gamma subunits**

  i. **Pituitary giantism/gigantism (as child) or acromegaly (as adult):** Too much production of growth hormone. Single amino acid substitution in the alpha unit of one of the G-proteins that regulates growth hormone.

  ii. **Cholera (due to *Vibrio cholerae*):** A disease that causes profuse diarrhea. It is caused by a bacterium that releases toxins. These toxins interact with a G-protein that regulates secretion of fluid into intestine. Under the influence of bacteria, this turns on the system too much, and too much fluid is produced.

  iii. **G-protein coding sequences are often the site of oncogenes.**

b. **Chloride channels**

  i. **Cystic fibrosis:** A disease that involves production of a thick mucus in respiratory system and pancreatic ducts, etc. There is a flawed gene that codes for protein of Cl$^-$ channels, so a thick mucus is produced.

VIII. **Drugs**

a. **Enhance response to cAMP**

  i. **Theophylline (SLO-BID):** A phosphodiesterase inhibitor; increases the half life of cAMP; an asthma drug. The signal to bronchi to dilate is done by a cAMP mechanism, so bronchi dilate with SLO-BID.

  ii. **Caffeine** is a phosphodiesterase inhibitor.

  iii. **(MAXAIR) (SEREVENT):** Stimulate adenyl cyclase, so produce cAMP more rapidly; asthma treatment.

  iv. **These drugs tend to be nonspecific because cAMP is nonspecific (and used in many processes).**

IX. Water balance shifts

a. Two main issues: **volume** and **osmolarity** (concentration). Remember, solutes do not readily cross membranes, but water does.

  i. **If one has severe diarrhea** (if diarrhea is very close to concentration of extracellular fluid [ECF]):

  (1) What happens to ECF volume?

  (2) What happens to ECF osmolarity?

  (3) What happens to intracellular fluid (ICF) volume?

  (4) What happens to ICF osmolarity?

  ii. **If one eats eat a bag of potato chips** (remember how Na^+ gets into plasma):
  (1) What happens to ECF osmolarity?

  (2) What happens to ECF volume?

  (3) What happens to ICF volume?

  (4) What happens to ICF osmolarity?

SUGGESTED READINGS

1. Guyton, A.C., and Hall, J.E. 2011. Chapters 1, 2, and 4 in *Textbook of Medical Physiology*, 12th ed. Baltimore: Saunders.
2. Alberts, B., Johnson, A., Lewis, J., Raff, M., Roberts, K., and Walter, P. 2002. Chapters 10 and 11 in *Molecular Biology of the Cell*, 4th ed. New York: Garland Science.

CHAPTER 1 QUESTIONS

1. _____ The largest fluid compartment
   A. Is found within the cell
   B. Is the only compartment that communicates directly with the environment
   C. Contains phosphodiesterase
   D. All of these
   E. Two of these

2. _____ Urine is part of which fluid compartment?
   A. Intracellular   D. Transcellular
   B. Intercellular   E. None of these
   C. Plasma

3. _____ Phospholipids
   A. Form ionic channels   D. Two of these
   B. Contain fatty acids   E. None of these
   C. May be integral or peripheral

4. _____ Membrane proteins
   A. May be part of the glycocalyx   D. All of these
   B. May function as enzymes   E. Two of these
   C. May function as adhesion molecules

5. _____ The cAMP system is associated with
   A. Ionic channels   D. None of these
   B. Phospholipase C   E. Two of these
   C. Diacylglycerol

6. _____ Integrins
   A. Bind to extracellular proteins   D. Two of these
   B. May attach directly to the cytoskeleton E. All of these
   C. Are proteins

For questions 7–10, choose one of the following:
   A. Increases or is greater than
   B. Decreases or is less than
   C. Has no effect or is equal to

7. _____ The volume of the intracellular compartment, as compared with the volume of the extracellular compartment

8. _____ The effect of increasing phospholipase C on the amount of 5'AMP that is formed

9. _____ The effect of increasing the amount of MAXAIR on the subsequent amount of adenyl cyclase activity

10. _____ The amount of energy expended for diffusion, as compared with the amount expended for osmosis

CHAPTER 2

Membrane Physiology

I. **Excitable tissue: A tissue that can respond to a stimulus and electrically propagate that signal down the membrane.** Virtually all cells have membrane potential (a **voltage gradient** across the membrane).

II. **Unequal distribution of ions**
   a. **Extracellular/interstitial:** Na^+: 150   **Intracellular:** Na^+: 15
                                    K^+: 5 K^+: 150
                                    Cl^-: 110 Cl^-: 4
                                    A^-: 45 A^-: 161*

       i. *Too big to go through channels of membrane
       ii. Zero net charge on either side of membrane (in equilibrium in both fluid compartments). As K^+ tries to move out to equilibrate concentration, some will get out, but it will still have a force on it from the A^- inside the cell. This allows a very small amount to get across the membrane to develop a membrane potential. The amount of K^+ that gets across is very small chemically, yet very significant in a voltage sense. In other words, the amount or net charge per side has not changed in a chemically measurable way, yet there is a voltage generated across the membrane; the chemical amount changed per side is equivalent to "spitting into the ocean" (see suggested readings: Wright, 2004).
       iii. There is an unequal distribution of ions and of charges across membrane.

III. **Factors causing unequal distribution of ions**
   a. **Na^+/K^+ pump:** Actively excluding Na^+ and actively including K^+.
       i. Does not do this in equal ratios; a 3 to 2 ratio (3 Na^+ out for every 2 K^+ in)
       ii. Actively taking Na^+ out and actively transporting K^+ in creates an unequal distribution of ions.
   b. **Donnan Equilibrium:** When a system has a large charged particle unable to move through the membrane, other members will have to be distributed unequally in order to reach equilibrium.

IV. Consequence of unequal distribution: causes a resting membrane potential

a. Unequal distribution of ions causes **resting membrane potential (RMP).** Body is at rest (or that nerve is not transmitting information). This voltage difference, mentioned above, is the RMP.

   i. RMP is due largely to diffusion of K^+.

b. K^+ crosses the membrane with the concentration/diffusion gradient.

   i. There is a potential difference across the membrane because of unequal distribution of K^+ on the outside and large anions on the inside.

   ii. **A voltage can exist across the membrane.**

   iii. At some point those forces are going to be in equilibrium.

c. **Goldman Equation**

   i. Accounts for permeability and concentration of all ions

   ii. Calculate the resting membrane potential (RMP)

   (1) Example:

   $EMF = (RT/F)\ ln\ (PK^+[K^+]out)/(PK^+[K^+]in) + (PNa^+[Na^+]out)/(PNa^+[Na^+]in) + (PCl^-[Cl^-]in)/(PCl^-[Cl^-]out)$

   (2) In this case P refers to permeability; the full equation accounts for all the major ions that contribute to the RMP.

d. **Nernst Equation**

   i. Accounts for the major player, K^+

   ii. $EMF = -61\ log\ [K^+]in/[K^+]out$

   (1) As $[K^+]out$ gets larger, EMF gets smaller (closer to 0) and vice versa.

   (2) If the two numbers become equal, then there is no difference ($log\ 1 = 0$).

   iii. Solve for a typical cell:

   (1) **What happens if you increase extracellular K^+?**

   (2) **What happens if you increase intracellular K^+?**

   iv. Resting membrane potential is **not totally** but **mainly** due to K^+.

   v. RMP is due to diffusion of K^+.

V. Action potential

a. Resting potential → Depolarization → Repolarization → Resting membrane potential

   i. Action potential composed of **two phases:** depolarization and repolarization

   (1) **Depolarization:** Na^+ is in high concentration outside but then the gates open, the permeability to Na^+ increases, and Na^+ rushes in to depolarize the membrane.

   (2) **Repolarization:** gates close and Na^+ permeability decreases. Also, K^+ permeability increases. K^+ goes out and carries positive charges with it to reestablish equilibrium.

   (3) **Resting membrane potential:** K^+ predominantly responsible, but Na^+ is predominantly responsible for action potential.

   (4) Remember, actual chemical equivalents that are moving are quite small!

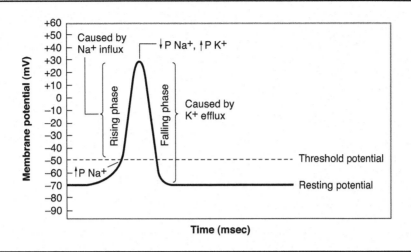

FIGURE 2.1

b. **Ionic explanation**

    i. A membrane's permeability to Na^+ increases (the gates open) and Na^+ rushes in to cause depolarization. During repolarization, K^+ doors reopen and Na^+ doors close. **K^+ is responsible for resting membrane potential; Na^+ is responsible for action potential.**

VI. **Factors causing action potentials: Factors that increase membrane permeability to Na^+**

  a. **Mechanical stimulation**

    i. The sciatic nerve innervates the gastrocnemius muscle of frog (making it contract). Frog's legs will twitch from mechanically stimulating nerve (and disturbing fluid mosaic enough to cause a mechanical stimulation). Na^+ rushes in and causes depolarization (causing muscle to contract).

  b. **Electrical stimulation**

    i. Experiment: Pinch nerve, and muscle contracts. We cannot quantitatively measure this; but with electrical stimulation, we can.

      (1) Advantage: quantitative/measure electrical stimulation

      (2) Components:

        a. How much? **Intensity** of stimulation is measured in volts.

        b. How long? **Duration** is usually measured in units of milliseconds.

        c. How often? **Frequency** of stimulation is measured in impulses per second.

        d. Electrical stimulators control all components.

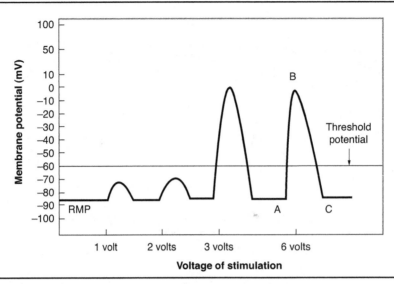

FIGURE 2.2

e. **Impulse has to reach threshold potential to become an action potential—"all or none".**
f. Have to overcome **excitability threshold** (the difference between the threshold potential and the resting membrane potential)
g. If a stimulus is insufficient to overcome the excitability threshold, it is called a subthreshold stimulus.
h. **Action potentials are "all or none" (3 volts or 6 volts yields the same result).**
i. If we stimulate immediately after the first stimulus, it will have no effect (between A and B—that is the **absolute refractory period [no action potential possible]**). If we stimulate between B and C, it takes a super stimulus to stimulate (**this is the relative refractory period).**
     (1) What is the maximum frequency we could transmit information in a neuron?
c. **Chemical stimulus (neurotransmitters)**
  i. Vocabulary of neurotransmitters
    (1) Neurons are single cells that make up a bundle called a nerve.
      a. Neurons have a long process called an axon.
      b. Neurons come in contact with another excitable tissue.
    (2) Synapse is the location where the neuron comes in contact with another excitable tissue.
      a. Information in this system flows one way down the axon (toward dendrites), from presynaptic structure to postsynaptic structure.
  ii. **Characteristics**
    (1) A neurotransmitter has to be **synthesized in the presynaptic structure.**
    (2) Action potential in presynaptic structure causes release of neurotransmitter into synaptic space and it then subsequently binds to a specific receptor on the postsynaptic structure.
    (3) Artificial application of the substance to the postsynaptic structure should cause the same response in the postsynaptic structure as was observed when stimulating the presynaptic structure.
    (4) A mechanism to terminate its action

VII. Factors altering excitability of membranes
  a. **Extracellular K$^+$**
     i. If we increase the concentration of extracellular K$^+$, then **excitability of membrane increases. The membrane has become hypopolarized and closer to threshold.**
  b. The separation between ions (due to small concentration gradient) causes voltage across membrane.
     i. The greater the concentration gradient, the more likely that K$^+$ will move out. The less the concentration gradient, the less likely K$^+$ will move to the extracellular space.
     ii. Nernst Equation: EMF = -61 log [K$^+$]in/[K$^+$]out
        (1) If we add extracellular K$^+$, then the [K$^+$]out increases, which means our resting potential line moves up and the magnitude of the excitability threshold gets smaller.
        (2) If [K$^+$]in gets smaller, EMF becomes less negative and the magnitude of the excitability threshold decreases.
  c. **Extracellular Ca^{++}**
     i. Increase in extracellular Ca^{++} decreases the excitability of the membrane. It does not **decrease EMF by altering RMP or TP.**
        (1) It does it mechanically: Ca^{++} blocks pores when its concentration is increased. When Ca^{++} is removed, Na$^+$ is able to move in easier.
     ii. **The body watches Ca^{++} and K$^+$ levels closely because they are so fundamental.**
  d. **Anesthetics:** Local anesthetics work by stabilizing membranes. Gates and channels cannot open or close, and thus we cannot propagate action potentials very well.
     i. Pain is only an action potential arriving at a certain place in the brain. With local anesthetics **(e.g., lidocaine)**, the action potential is not getting to the brain to register pain.

VIII. Drugs
  a. **Tetrodotoxin (TTX):** in puffer fish; blocks Na$^+$ channels.
  b. **Saxitoxin (STX):** found in marine invertebrates; blocks Na$^+$ channels. Reddish color is responsible for "red tide."
  c. **Dendrotoxin:** blocks K$^+$ channels; makes it hard to repolarize. Found in mamba snakes.

SUGGESTED READINGS

1. Guyton, A.C., and Hall, J.E. 2011. Chapter 5 in *Textbook of Medical Physiology*, 12th ed. Baltimore: Saunders.
2. Wright, S.H. 2004. Generation of resting membrane potential. *Adv. Physiol. Educ.* 28:139–42.
3. Koeppen, B.M. and Stanton, B.A. 2008. Chapters 1, 2, and 3 in *Berne & Levy Physiology*, 6th ed. St. Louis: Mosby.

CHAPTER 2 QUESTIONS

1. _____ During depolarization
   A. The permeability of the membrane to potassium is the same as it is during rest
   B. Sodium channels are closed
   C. The polarity is reversed
   D. Two of these
   E. All of these

2. _____ A sub-threshold stimulus
   A. Causes an action potential
   B. Can alter sodium permeability
   C. Decreases threshold potential

   D. All of these
   E. Two of these

3. _____ During depolarization
   A. The membrane is absolute refractory
   B. Sodium gates are open
   C. Potassium gates open

   D. All of these
   E. Two of these

For questions 4–7, choose one of the following:
   A. Increases or is greater than
   B. Decreases or is less than
   C. Has no effect or is equal to

4. _____ The effect of increasing intracellular potassium on the magnitude of the excitability threshold

5. _____ The effect of increasing a membrane's permeability to potassium on the polarity of the membrane

6. _____ The effect of increasing extracellular calcium concentration on the magnitude of the threshold potential

7. _____ The effect of saxitoxin the permeability of sodium channels

8. A large amount of saturated KCl solution is injected into an animal. What happens to the animal's heart? Please explain why in the space below.

CHAPTER 3

Muscle Physiology

I. Introduction
   a. **Types and Classification**
      i. Control mode
         (1) **Voluntary:** Make muscle contract when you want it to—**skeletal**
         (2) **Involuntary:** Unable to voluntarily control—**cardiac and visceral**
      ii. Functional types of muscle
         (1) **Skeletal muscle:** moves limbs and body; under voluntary control
         (2) **Cardiac muscle:** under involuntary control
         (3) **Visceral muscle:** under involuntary control
      iii. Histological classification
         (1) **Striated** (under a microscope, it looks striped)
            a. Skeletal muscle
            b. Cardiac muscle
         (2) **Smooth**
            a. Visceral muscle
   b. **Skeletal Muscle Properties**
      i. **Irritable:** There exists a resting membrane potential across membrane.
      ii. **Propagate:** Can induce an action potential and propagate it across the cell membrane
      iii. **Tension:** Capable of developing tension
         (1) The average male would be capable of ~25 tons of tension if all his muscles were contracted at once.
      iv. **How do skeletal muscles contract?**
         (1) Contract in response to the nervous system stimuli

c. **Motor Unit**
   i. The relationship between the nervous system and the muscle it innervates
      (1) A motor unit is a single neuron and all the muscle fibers it innervates.
   ii. **How many muscle fibers to a motor unit?**
      (1) For precise control over muscles (such as in fingers), there are typically **very few muscle fibers to a motor unit**.
      (2) For larger, gross movements, it is possible for there to be **thousands of muscle fibers per motor unit**.
d. **Vocabulary**
   i. **Muscle**: A muscle is composed of many, many muscle cells/fibers.
   ii. **Muscle fiber:** A single cell
   iii. **Neuron:** A single cell that is branched at its ends and innervates skeletal muscle fibers
   iv. **Nerve:** A bundle of neurons

II. **The Neuromuscular Junction**: The place at which a neuron comes in close contact with a muscle fiber. There is a space between the two!

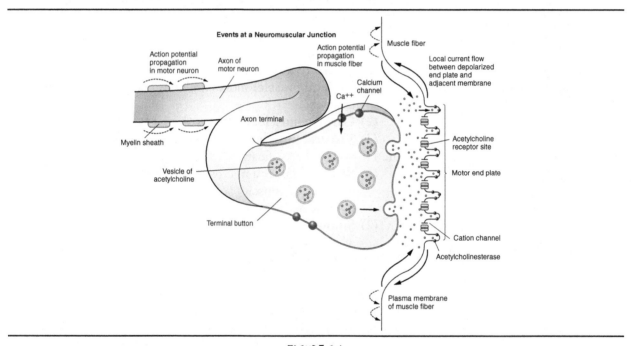

FIGURE 3.1

a. **Anatomy**
   i. There is no continuity between the junction at the end of the neuron and the muscle fiber; **they do not touch**—there is a space between the two.
   ii. Muscle fibers contract in response to being stimulated by nervous system.
      (1) An action potential in the neuron causes muscle fibers to contract.
   iii. **Neuronal side**
      (1) Vessels/vesicles containing **acetylcholine (ACH,** a neurotransmitter)
      (2) **Many mitochondria in the area, due to the high level of metabolic activity**

iv. **Muscle fiber side**
  (1) Tends to be **highly convoluted** (or wrinkled) in order to increase surface area
  (2) **Acetylcholine receptor** sites bind acetylcholine and make conformational changes in the muscle fiber cell membrane.
  (3) The enzyme **acetylcholinesterase breaks down acetylcholine.** This enzyme breaks down free ACH. Free ACH is in equilibrium with bound ACH, so breaking down the free stuff causes ACH to come off of receptors.

b. **Physiology**
  i. An **action potential is propagated down the neuron** to the terminal end. This causes **Ca^{++} to flow into the neuron** from the interstitial fluid.
  ii. This causes the **vesicles** that contain acetylcholine to **migrate up to the membrane.**
    (1) **How? Still under investigation, but…**
      a. Action potential comes down, causes **synaptobrevin** (on vesicle membrane) to become involved with **syntaxin** and **SNAP25** (both on membrane of neuron). Ca^{++} causes the interlocking of these three proteins—the membrane proteins pull the vesicle toward the neuronal membrane and "dock" it. **Ca^{++}** then comes and interacts with the **synaptotagmin** on the vesicle membrane, causing the vesicle to exocytose ACH into the neuromuscular junction space.
    (2) Vesicles release or exocytose acetylcholine into the space between the muscle fiber and the neuron.
      a. Diffusion distance is narrow.
  iii. There is a **high density of acetylcholine receptors on the muscle fiber.**
    (1) Acetylcholine causes the receptors (ACH-Rs) to change shape/conformation, which directly allows Na$^+$ to enter the cell (ACH-R is a Na$^+$ channel) at the area of the muscle we call the motor end plate (MEP). This generates an "end plate potential" (EPP), which in turn causes a "regular" depolarization of the adjacent membrane. This potential is then propagated and involves the "normal" action potential generation (voltage-gated Na$^+$ and K$^+$ channels, etc).
  iv. This increases the Na$^+$ permeability of the muscle fiber. As Na$^+$ rushes into the muscle fiber, **depolarization of the sarcolemma** occurs.

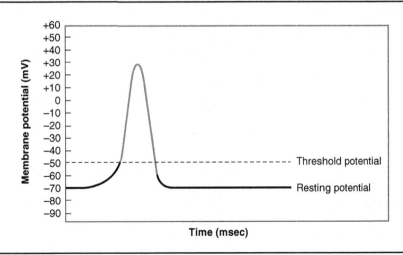

FIGURE 3.2

(1) The neurotransmitter acetylcholine is used to transfer the action potential to the muscle fiber.

    v. **Acetylcholinesterase** breaks down acetylcholine and deactivates it, so **the signal begins to terminate**.

    vi. **Acetylcholine transfers an action potential** from the neuron to the muscle fiber.

III. The Muscle Fiber

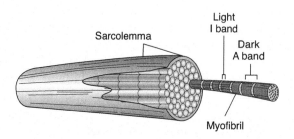

FIGURE 3.3

a. Introduction

    i. Muscle fibers are **multinucleated** and contain **many mitochondria (if type I)**.

      (1) They are covered by a cell membrane, or **sarcolemma**.

    ii. Each skeletal muscle fiber is innervated by its own neuron at the neuromuscular junction/motor end plate.

b. **Myofibrils, Filaments, Actin & Myosin, Titin**

    i. **Myofibrils:** Muscle fibers are composed of myofibrils. These look like straws that are together underneath one sarcolemma.

      (1) They are made up of subunits/filaments. Within each myofibril are thick filaments, thin filaments, and titin.

      (2) If you look at myofibrils longitudinally, you can see that the thick filaments, thin filaments, and titin are interconnected, that they partially overlap one another, and that the sections repeat themselves.

      (3) **The Z line** is perpendicular to the thin filaments.

      (4) The part of the myofibril between the two Z lines is a **sarcomere—"the functional unit."**

      (5) **The portion of the myofibril that contains all of the thick filaments** and the ends of the thin filaments (or where the thick and thin overlap) is **the A band**.

      (6) Just the thin filament and titin is the **I band**.

      (7) Within the A band, the part that contains **only thick filament** is the **H zone**.

      (8) **The M-line** links the thick filaments together.

      (9) The protrusions on the thick filaments are the **cross-bridges**. These will interact with thin filaments during muscle contraction.

    ii. The thick filaments are attached to the Z line by the filament called **titin**.

      (1) Titin stabilizes the thin and thick filaments.

      (2) Titin also holds the thick filament appropriately in the myofibril.

    iii. **Nebulin:** Spans the length of the entire thin filament; thought to measure correct length of filament and appropriately hold the thin filament.

    iv. Dystrophin: Anchors entire myofibril array to cell membrane.
    v. **The thin filaments** are primarily composed of actin and are sometimes referred to as "actin filaments" (but they also have troponin and tropomyosin).
        (1) Beads = **G-actin** ("globular")
        (2) Strands = **F-actin** ("filament"; the helical structure with myosin binding sites)
    vi. **The thick filaments** are primarily composed of myosin and are sometimes referred to as "myosin filaments."
    vii. There are "stripes" on the skeletal muscle because of this **regular arrangement between thick filaments and thin filaments** (sarcomere).

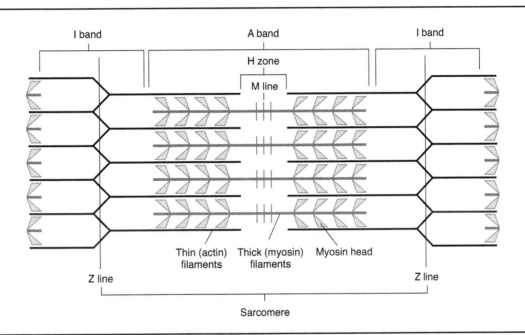

FIGURE 3.4

  c. **Membranous Systems:** Part of each muscle fiber
    i. **Sarcoplasmic Reticulum (SR):** A type of specialized endoplasmic reticulum in skeletal muscle.
        (1) They are dumbbell-shaped structures that run parallel to muscle fiber.
            a. Narrow part: Longitudinal portion
            b. End round part: Lateral sacs
        (2) It embraces the myofibrils (or **wraps around them**).
        (3) The role of the SR relates to the Ca^{++} movement within the muscle fiber.
            a. **Ca^{++}** is stored in the SR.
            b. **Ca^{++}** can be released from the SR, and it can also be reclaimed by it.
    ii. **T-tubular System**
        (1) **Runs perpendicular to the muscle fiber** and is essentially many invaginations of the membrane
        (2) T-tubules are "pushed" into the sarcolemma.
            a. Like pushing your finger into a balloon (draw this)

(3) What kind of fluid is found in the T-tubules?

(4) What kind of fluid is found in the SR?

(5) The T-tubules help to propagate action potentials deep into the muscle fiber.

iii. Sarcoplasmic reticulum and the T-tubular system: "the triad"

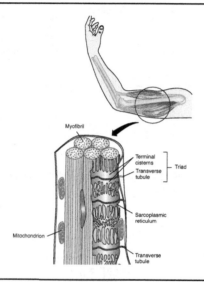

FIGURE 3.5

(1) The SR runs in the same plane as the muscle fibers.
(2) The T-tubules are interspersed **between the ends of the "dumbbells," or lateral sacs, of the SR.**
   a. **Triads:** Two lateral sacs of the SR and the cross-section of the T-tubule.
(3) **T-tubules propagate action potentials (because they are part of the sarcolemma)** up to the junction of the SR.
   a. The **action potential travels down the T-tubule,** causing a **conformational change in DHP receptors** (part of T-tubule membrane); this in turn "uncorks" the **ryanodine receptors on the SR** to allow **Ca^{++} out of the SR into the cytoplasm of the cell.**

IV. The Thick Filament
  a. Structure
    i. Contains only myosin
    ii. The myosin molecule behaves as if it were hinged.

    iii. The "head" bends in when the muscle contracts.
    iv. Head has a site for:
       (1) Binding to the actin molecule,
       (2) Binding ATP, and
       (3) ATPase activity.
  b. Characteristics
    i. Binds and hydrolyzes ATP. ATP is needed to release binding between myosin and actin. ADP and Pi are stored on head in the relaxed state. These are released as the energy is used from the hydrolysis in a contraction. ATP is then again used to relax.

V. The Thin Filament
  a. Structure

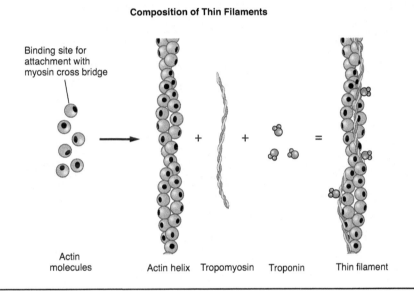

Composition of Thin Filaments

Binding site for attachment with myosin cross bridge

Actin molecules   Actin helix   Tropomyosin   Troponin   Thin filament

FIGURE 3.6

    i. A thin filament, arranged in a helix:
       (1) Like the beads of a necklace
       (2) G-actin (monomers) and **F-actin** (strands)
    ii. Tropomyosin runs along (lays in the groove of) the actin filament.
       (1) Discontinuous every 7 or so beads.
    iii. Troponin is at the end of each tropomyosin.
       (1) Globular—has three components:
          a. TnT: binds to tropomyosin
          b. TnI: binds it to actin
          c. TnC: binds Ca^{++}
       (2) Troponin is bound to tropomyosin and to actin.
       (3) Also has the binding site for Ca^{++}
       (4) Blocks the binding site on actin for myosin.

    iv. Actin is essentially two strips of beads that are twisted into a helix, with tropomyosin and troponin running along it as described.

  b. Relationship between actin & myosin filament
    i. Actin and myosin partially overlap each other.
    ii. Cross-bridges:

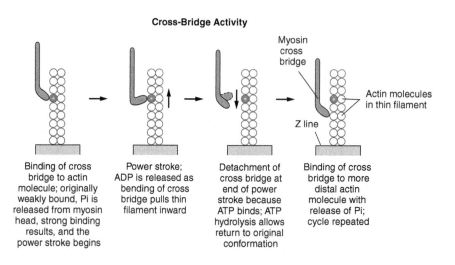

Cross-Bridge Activity

Myosin cross bridge

Actin molecules in thin filament

Z line

| Binding of cross bridge to actin molecule; originally weakly bound, Pi is released from myosin head, strong binding results, and the power stroke begins | Power stroke; ADP is released as bending of cross bridge pulls thin filament inward | Detachment of cross bridge at end of power stroke because ATP binds; ATP hydrolysis allows return to original conformation | Binding of cross bridge to more distal actin molecule with release of Pi; cycle repeated |

FIGURE 3.7

    iii. The cross-bridges on the heads of the myosin interact with actin.
      (1) They bind actin, bend in, pop up, bind again, bend in, pop up...
      (2) The myosin head attaches to actin and pulls actin farther into the H zone.
        a. The muscle fibers shorten, so the muscle shortens.

VI. The Physiology of Contraction: Ratchet Theory
  a. The process by which muscles contract.
  b. What actually gets shorter during contraction?
    i. Muscle?

    ii. Muscle fibers?

    iii. Myofibrils?

    iv. Thick or thin filaments?

      v. H-zone?

      vi. A band?

      vii. I band?

      viii. Sarcomeres?

VII. **Regulation of Contraction:** In order for muscles to contract, we have to get actin and myosin together. **Three things regulate: Troponin/Tropomyosin, Ca^{++}, and ATP.**
  a. **Regulation by Troponin & Tropomyosin**
    i. Example: Purified filaments of actin stripped of troponin and tropomyosin will bind to myosin and contract.
    ii. Actin and myosin cannot get together, since troponin and tropomyosin are in the way (**tropomyosin lays in the groove, blocking the myosin binding sites of the actin**).
      (1) **Troponin and tropomyosin inhibit** actin and myosin.
      (2) In order for actin and myosin to get together, cross-bridges on myosin must get to the binding sites on actin.
      (3) We have to move tropomyosin in order for actin and myosin to interact.
        a. **Ca^{++} will bind to troponin and change the conformation of it so that it moves tropomyosin off of the myosin binding site** (it actually moves deeper into the filament, out of the way). The inhibition of actin and myosin binding is relieved.
        b. **Ca^{++} allows myosin and actin to bind.**
  b. **Regulation by Calcium**
    i. **Ca^{++} is typically stored in the SR** (specialized type of endoplasmic reticulum).
      (1) Intracellular Ca^{++} in the SR
    ii. Something has to cause the release of Ca^{++} from within the SR.
      (1) If the Ca^{++} is no longer in the SR, it is free in the intracellular fluid.
      (2) Now, it is free to bind to troponin.
    iii. The SR also regulates the taking back up of Ca^{++}. It is pumped actively back into the SR.
      (1) **The Ca^{++} leaves the troponin,** and so troponin changes conformation. The muscle relaxes as ATP relaxes the strong binding of actin/myosin and tropomyosin moves back into the groove to block the myosin binding site.
    iv. **Ca^{++} is released in response to an action potential:**
    v. **How?** The T-tubules have proteins called **DHP receptors,** and the SRs, or lateral sacs, have proteins called **ryanodine receptors.**
      (1) **Ryanodine receptors** exist in groups of four and are integral proteins.
      (2) **The DHP receptor is right up against the ryanodine receptors.** When the action potential is propagated down the T-tubules, it causes the DHP receptors to change conformation and "uncork" the holes on the SR, allowing Ca^{++} to be released.

(3) This allows Ca^{++} to come through the channels.
    a. Depolarization causes this.
    b. Then the DHP receptor changes its conformation back and stops the Ca^{++} movement.
  vi. **Excitation Coupling**
    (1) The coupling of the electrical event (action potential) to the mechanical event (muscle shortening or contraction)

c. **Regulation by ATP**
  i. Must have ATP for muscles to contract.
  ii. Myosin binds ATP and hydrolyzes it, and the energy is stored in the head of myosin.
    (1) Example: Like a cocked gun
  iii. There is no interaction between the head of myosin and actin because of troponin and tropomyosin.
    (1) Ca^{++} binds to troponin and enables myosin to bind to actin.
    (2) In order to make the myosin head bend according to the ratchet theory, **energy is released.**
      a. As long as we have plenty of ATP, the head pops up again and binds to actin over and over (bends in, pops up, bends in, pops up).
      b. **ATP is needed to relax a myosin-actin binding (energy stored at the head, etc.).**
      c. All this time the filament is getting shorter.
      d. It contracts until the Ca^{++} is taken back up by the SR.
      e. If there is no Ca^{++} left, then the filament relaxes.
      f. **Scenarios**
        i. **Plenty of ATP, plenty of Ca^{++}?**

        ii. **Plenty of ATP, but no free Ca^{++}?**

        iii. **Plenty of Ca^{++}, but no ATP?**

VIII. Energy Sources: How do we generate ATP?
  a. **Creatine phosphate or "phosphocreatine (PCr)" and "The Immediate Energy System"**
    i. PCr is a high-energy source like ATP.
    ii. We use PCr to immediately phosphorylate ADP back to ATP.

$$\textbf{PCr + ADP (creatine kinase)} \rightarrow \textbf{Creatine + ATP}$$

    iii. So **muscles store PCr** in order to rapidly replenish ATP.
  b. **Glycolysis**
    i. An "anaerobic" process (can run without O_2; of course, also runs when O_2 is present)
    ii. In cytosol
    iii. Produces ATP quickly, yet not a large yield
  c. **Oxidative phosphorylation**
    i. Must have a large supply of oxygen
    ii. In mitochondria
    iii. Yields large amounts of ATP, but slowly

IX. Skeletal Fiber Types & Characteristics
a. **Type I:** "Slow Oxidative"
   i.   These **contract slowly and relax slowly.**
   ii. Oxidative means **many mitochondria.**
   iii. These muscle fibers contain the pigment myoglobin, which stores oxygen (has **a reddish color**).
   iv. Resistant to fatigue
   v.   Predominance in anti-gravity muscles, such as in the back or buttocks
b. **Type IIa:** "Fast Oxidative"
   i.   Contract and relax quickly
   ii. Fairly resistant to fatigue
   iii. Some mitochondria
   iv. Use much ATP
   v.   Contain some myoglobin
   vi. Example: In the breast of migratory birds
c. **Type IIb:** "Fast Glycolytic"
   i.   Very fast contractions
   ii. Fatigue easily
   iii. Not many mitochondria
   iv. White appearance; few O_2 binding elements
   v.   Depending on the usage of the term, "Type IIb" can refer to the muscle type or the myosin heavy chain (MHC). MHCs in humans are actually type IIx; the IIb isoform was discovered in rats. **It is highly recommended** that you read the excellent article in the suggested readings entitled "Human skeletal muscle fiber type classifications." This should clear up any confusion and makes the point that there are many different ways to classify any tissue (based on twitch speed, oxidative enzyme activity, etc.).

X. Form Curves: The simple twitch of one muscle fiber:

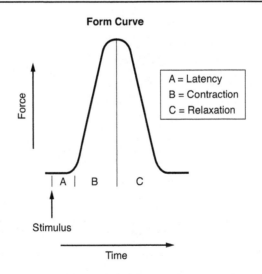

FIGURE 3.8

Is this "all or none"?

Why or why not?

What happens from A to B?

In smooth muscle, excitation coupling occurs via a phospholipase C mechanism.

In cardiac muscle?

What is the relationship between the force of contraction and the amount of Ca^{++} released into myofibril in cardiac and skeletal muscles?

XI. Smoothing and Grading
  a. Multiunit or multiple motor summation

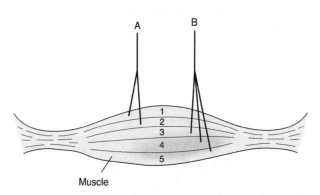

Above, motor unit A consists of the motor neuron and all the muscle fibers it innervates (1 and 2). Motor unit B would be that motor neuron and all the muscle fibers it innervates (3, 4, and 5). Note that an action potential down any motor neuron contracts ALL of the muscle fibers innervated by it. Thus, muscle fibers are typically of the same type if they are innervated by a single neuron. Also, notice that these muscle fibers are just part of a "muscle," like the biceps brachii.

FIGURE 3.9

Above, each motor unit innervates different muscle fibers.

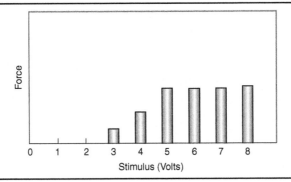

FIGURE 3.10

(1) At _____ volts it has reached **threshold for at least one motor unit.**
(2) Does this produce a **graded response** or **an all-or-none response?**
(3) Reached threshold for all motor units at _____ volts
(4) We recruit smaller motor units first ("size principle").

ii. **Graded response involves recruiting different numbers of motor units in a muscle.**
   (1) If we have very good control of a muscle, then there are **only a few muscle fibers per motor unit.**
   (2) **If we have many muscle fibers per motor unit**, then we have limited control.
b. **Asynchronous Firing**
   i. A muscle with 1,000 motor units has the capability of 1,000 factorial (!) combinations of movements.
   ii. Through motor learning, our brains learn to recognize different ways to move or work our motor units.
      (1) The brain recognizes "heavy" or "light" things.
c. **Wave Summation**
   i. Alters frequency (number of impulses per second) instead of voltage
      (1) Responses are smoother and smoother, and also more graded.

      (2) Point where muscle stays contracted is called "tetany."
         a. The persistence of free calcium in the sarcoplasm causes the muscle to remain continuously contracted.

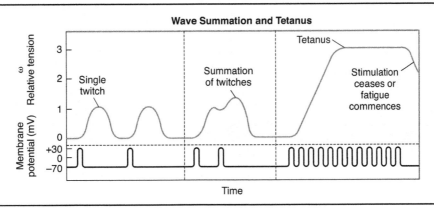

FIGURE 3.11

d. **Starling's Law:**
    i.   **Within physiological limits, if you increase the resting length of the muscle then you increase the force of contraction.**
        (1) By stretching the muscle, you expose more binding sites on the actin and therefore increase the force of contraction.

XII. **Smooth Muscle**
  a. **Structure**
    i.   Has a single nucleus
    ii. The cells are close together.
    iii. They are not striated, because there is no regular interaction between actin and myosin.
    iv. The **SR is far less developed than skeletal.**
    v. There are **no T-tubules.**
    vi. Made up of actin, myosin, and tropomyosin but **no troponin**
    vii. **Has an intermediate filament that is not involved in the contraction but is thought to be part of the cytoskeleton** or structure of the smooth muscle cell
    viii. There is no regular arrangement between actin and myosin.
  b. **Organization:** smooth muscle is organized two ways.
    i.   **Multiunit:** Each smooth muscle cell is privately innervated (like skeletal muscle).
        (1) Examples: The iris of the eye, blood vessels, hair follicles
    ii. **Single Unit/Unitary/Visceral: A functional syncytium** (operates as one)
        (1) Smooth muscles are together in very large sheets (and close to one another in these sheets).
        (2) The action potential occurs in any smooth muscle cell and is propagated to its neighbor (and then to its neighbor) via connexons.
        (3) Not privately innervated
  c. **Physiology** (Actin-linked regulation vs. **Myosin-linked regulation**)
    i.   **Smooth muscle is myosin-linked.**
    ii. There is an increase in the intracellular Ca^{++} from either the extracellular fluid or the SR.
        (1) It then binds to calmodulin.
        (2) The Ca^{++}-calmodulin complex activates myosin kinase.
        (3) The myosin kinase phosphorylates myosin.
        (4) The myosin is now activated and can interact with actin.

XIII. **Pathologies**
  a. **Muscular Dystrophy**
    i.   An inherited disease that produces progressive muscle weakness and deteriorates the muscle tissue.
        (1) The muscle tissue degenerates and is replaced by fatty and fibrous tissue.
    ii. **Duchene's Muscular Dystrophy (DMD)** Occurs in young males (X-linked recessive).
        (1) Degenerates skeletal muscle first and then facial muscles and cardiac muscles.
        (2) By age 12, patients usually are nonambulatory (cannot walk).
        (3) Often patients **do not survive past early 20s.**
        (4) The product of the DMD gene (where there is mutation) is the protein *dystrophin.*
          a. **People suffering from DMD lack functional dystrophin protein.**
          b. Normally dystrophin anchors myofibrillar apparatus to sarcolemma.

b. **Functional Dystonias**
   i. Focal dystonias
     (1) Causes involuntary and sustained muscle contractions
     (2) Could cause the twisting, turning, or squeezing of a body part
     (3) Examples: **In the neck; writer's cramp**
   ii. The problem is in the central nervous system.

c. **Myasthenia gravis**
   i. Receptors that are sensitive to acetylcholine on sarcolemma become progressively inactive. **They are attacked by one's own immune system.**
   ii. Causes progressive muscle weakness and problems with breathing (is **often fatal**).

d. **Malignant Hyperthermia**
   i. In response to anesthetics, this is when the **SR ryanodine receptors get "locked open"** and Ca^{++} is uncontrollably released.
   ii. If not treated, it results in hyperthermia, tachycardia, and/or muscle rigidity.

XIV. **Drugs**
  a. **Neuromuscular Junction**
    i. **Stimulators** (Acetylcholinesterase Inhibitors)
      (1) Work by the fact that they are acetylcholinesterase inhibitors.
        a. **They increase the half-life of acetylcholine.**
      (2) **Physostigmine (ESERINE)**
      (3) **Organophosphates:** Present in insecticides and very potent.
      (4) **War gases**
    ii. **Inhibitors**
      (1) **Curare (dtc, TUBADIL, D-tubocurare):** blocks the acetylcholine receptors.
        a. Curare attaches to the receptor but does not cause an action potential.
        b. **A "nondepolarizing blocker"**
        c. Is naturally occurring in some plants
        d. Works postsynaptically
      (2) **Succinylcholine (ANECTINE):** binds to receptors and allows one action potential, but then blocks it.
        a. **A "depolarizing blocker"**
        b. In the surgical setting, used for intubation
        c. The diaphragm could be poisoned if too much is given, so one must be careful.
      (3) **Botulin (BOTOX):** a toxin produced by *Clostridium botulinum* bacteria.
        a. A presynaptic blocker
        b. It blocks the release of acetylcholine from the neuron by **damaging release mechanism.**
        c. Only minute amounts are used.
        d. Botox is used to relax an overly contracting muscle or to smooth out wrinkles.
        e. Eventually wears off (BOTOX); used to treat functional dystonias.
      (4) **Tetanus:** toxoid produced by *Clostridium tetani* bacteria.
        a. **Interferes with GABA secretion that inhibits in pathways to muscle contraction (inhibitory neurons)**
        b. **Causes too much excitement**
        c. Spastic contractions

b. Muscle Relaxants
   i. These work centrally!
      (1) (FLEXERIL)
      (2) (SKELAXIN)
      (3) (SOMA)

SUGGESTED READINGS

1. Scott, W., Stephens, J., Binder-Macleod, S.A. 2001. Human skeletal muscle fiber type classifications. *Phys. Ther.* 81:1810–6.
2. Brooks, G.A., Fahey, T.D., and Baldwin, K.M. 2005. Chapters 17, 19, and 20 in *Exercise Physiology: Human Bioenergetics and its Applications*, 4th ed. New York: McGraw-Hill.
3. Guyton, A.C., and Hall, J.E. 2011. Chapters 6–8 in *Textbook of medical physiology*, 12th ed. Baltimore: Saunders.
4. Koeppen, B.M. and Stanton, B.A. 2008. Chapters 12–14 in *Berne & Levy Physiology*, 6th ed. St. Louis: Mosby.

CHAPTER 3 QUESTIONS

1. _____ The number of muscle fibers in a motor unit supplying a muscle in your back is (A) greater than, (B) less than, or (C) equal to the number supplying a muscle in your finger.

2. _____ is that part of the A band that contains only the thick filaments.

3. _____ The following structures contain intercellular fluid:
   A. T-tubules   D. All of these
   B. Sarcoplasmic reticulum   E. Two of these
   C. Muscle fiber

In questions 4–10, choose from the following to represent the various parts of a form curve of a skeletal muscle being directly stimulated by an electrical stimulus:
   A. Latency
   B. Contraction
   C. Relaxation
   D. None of these; false

4. _____ Action potential travels down the t-tubule.

5. _____ Ca^{++} moves out of SR.

6. _____ Ca^{++} moves in from outside the cell.

7. _____ Ca^{++} binds tropomyosin.

8. _____ ATP binds actin.

9. _____ Ca^{++} is sequestered back into the SR.

10. _____ ATP is first hydrolyzed.

CHAPTER 4

Neurophysiology I

I. Introduction
   a. **Comparison of nervous and endocrine systems**
      i. Both systems integrate various physiological systems, but the **nervous system is wired by neurons,** whereas the **endocrine system is a chemical system that uses hormones.**
      ii. Response time of nervous system is in **milliseconds**; endocrine system **may take up to days.**
         (1) It is like the telephone (nervous system) vs. the postal service (the endocrine system).
      iii. The nervous response is short lived.
         The endocrine response is long lived.
   b. **Components of reflex arc:** The simplest way in which the nervous system works is via a reflex!
      i. A **receptor** (not a protein molecule, like we used this word before!) that detects some threat to homeostasis.
      ii. The **afferent (sensory) neuron** takes the information from the receptor to the integrating circuit.
      iii. **The integrating circuit processes it.**
      iv. The **efferent (motor) neuron** receives information from the integrating circuit and takes information to the effector.
      v. **The effector** works to restore homeostasis.
      vi. **The effector is one of these four:**
         (1) Skeletal muscle
         (2) Smooth muscle
         (3) Cardiac muscle
         (4) Glandular structure

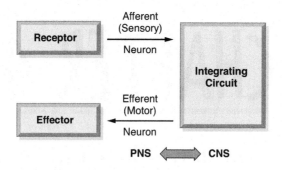

FIGURE 4.1

II. Organization of nervous system
    a. Anatomical Organization
        i. Organized into the **central nervous system** (CNS) and the **peripheral nervous system** (PNS)
            (1) The CNS is composed of the **brain** and the **spinal cord.**
            (2) The PNS is composed of the **cranial nerves** (nerves that arise/terminate in the brain) and **spinal nerves** (nerves that arise/terminate in the spinal cord).
                - Cranial nerves typically innervate structures of the head and neck, whereas spinal nerves innervate all parts of the body.
    b. Functional Organization
        i. Organized into the **somatic nervous system** and the **visceral nervous system**
            (1) The somatic nervous system is associated with the functioning of the soma (body wall and/or limbs) and is generally voluntary.
                - The components of reflex are as described above.
                - The somatic effector is always skeletal muscle.
            (2) The visceral nervous system is associated with the functioning of the viscera (heart, gastrointestinal system, etc.) and generally is involuntary.
            (3) Another name for the visceral nervous system is the autonomic nervous system, which can be further subdivided into the **sympathetic and parasympathetic** nervous systems.
                -The components of a visceral reflex are illustrated below.
                -The visceral effector is always smooth muscle, cardiac muscle, or some glandular structure.

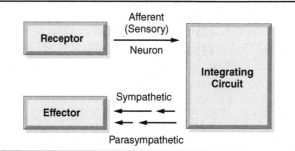

FIGURE 4.2

-Note that in the efferent component of the reflex there are two pathways between the integrating circuit and the effector, and in each of those two pathways there are two neurons.
-In one of these pathways (sympathetic) there is a short neuron and then a long neuron. In the other (parasympathetic) there is a long neuron and then a short one.
-Most of visceral effectors are **doubly innervated**—have both sympathetic and parasympathetic pathways innervating them.

c. The integrating circuit is always found within the CNS in one of three areas:
  i. Cerebrum: conscious sensation; somatic reflexes
  ii. Brainstem: unconscious sensation; somatic and visceral reflexes
  iii. Spinal cord: unconscious sensation; somatic and visceral reflexes
d. The efferent and afferent components are always found within the PNS.
e. Integration of the anatomical and functional organizations can be illustrated with four quadrants.

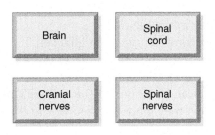

FIGURE 4.3

  i. Brain quadrant
  -The brain acts as the integrating circuit for both somatic and visceral reflexes.
  ii. The spinal cord quadrant
  -The spinal cord acts as the integrating circuit for both somatic and visceral reflexes
  iii. Cranial nerves quadrant
    (1) Cranial nerves have both a somatic (afferent and efferent) function as well as a visceral (afferent and efferent) function. Note: Their visceral function is limited to only parasympathetic function. Also, remember that cranial nerves mostly innervate structures of the head and neck.

    - What would be an example of a somatic efferent function governed by a cranial nerve? Blinking your eye would be an example. Can you think of another?

    -What would be an example of a parasympathetic afferent function governed by a cranial nerve? The sensation of nausea would be an example. Can you think of another?

  iv. Spinal nerves quadrant
    (1) Spinal nerves have somatic (afferent and efferent) functions, sympathetic (afferent and efferent) functions, and parasympathetic (afferent and efferent) functions.

-What would be an example of a somatic efferent function governed by a spinal nerve? Moving your hand would be an example. Can you think of another?

- What would be an example of a parasympathetic afferent function governed by a spinal nerve? The sensation of a full bladder would be an example. Can you think of another?

Note: If you are wondering how to tell if something is sympathetic or parasympathetic, don't sweat it. We will enable you to tell the difference a little later.

III. Cellular neurophysiology
   a. Cell types
      i. **Glial cells:** These do not function to transmit action potentials *per se*; that is the neurons' job. **Glial cells' main role is to support the neurons (although they may communicate with other glial cells, etc).**
         (1) **Astroglia:** Found only in the CNS
            a. They are the most common type of a glial cell.
            b. They are the **glue that holds the neuron in its correct spatial relationship.**
            c. They **establish the blood-brain barrier.**
            d. During embryological development the astroglia are **the scaffold by which the neurons learn to grow;** they guide the neurons to the right location.
            e. Astroglia **repair injuries** in neurons.
            f. They are **involved in nourishing neurons.**
            g. They also **uptake neurotransmitters.**
         (2) **Oligodendria:** found only in the CNS
            a. Within the CNS they **lay down myelin** on the neurons.
            b. They also **regulate pH.**
            c. They also **participate in iron metabolism.**
         (3) **Microglia:** These are phagocytotic. They endocytose various materials when there is damage to neurons from foreign debris or invaders.
            a. They phagocytose cytosolic debris.
            b. They mediate some immune responses.
            c. These may be overzealous and cause dementia.
         (4) **Ependymal cells:** These are found in the CNS, within brain cavities.
            a. They **produce CSF.**
            b. These are a possible source of neural stem cells.
         (5) **Schwann cells:** These function in the **PNS** and have a job similar to that of the oligodendria's function.
            a. They **lay down myelin on neurons.**
         (6) **Satellite cells: PNS**
            a. Surround the dorsal root ganglia (help protect dorsal root)
      ii. **Neuron**
         (1) Neurons **lack a mitotic apparatus,** whereas the glial cells retain their mitotic apparatus.
         (2) There are **more glial cells than neurons.**
         (3) If you have a tumor on neural cells it is typically the glial cells, not the neurons.
         (4) **Neurons are irritable,** which means they have a resting membrane potential and they can propagate an action potential.
         (5) At the terminal end of the neuron, neurotransmitters are released.
         (6) **Neuroblastoma: tumor of neural tissue, typically in pediatric adrenal glands**

b. **Neuron**
   i. **Functional characteristics**

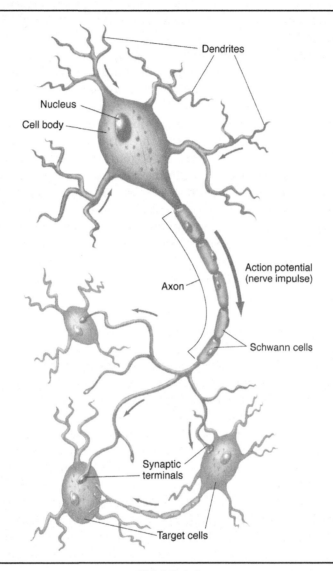

FIGURE 4.4

ii. **Anatomy**
   (1) Cell body: contains intracellular organelles as in other cells
      a. There is no mitotic apparatus, so cannot self-replicate
      b. Nuclei: These are nerve cell bodies located in the CNS. **Nucleus:** This is a single nerve cell body in the CNS.
      c. These nuclei are usually networked together to perform certain functions and are referred to as "centers."
      d. Ganglion: a nerve cell body located in the PNS. **Ganglia:** nerve cell bodies in the PNS.

(2) Dendrites: Multi-branched and similar to a nerve cell body
  a. These are more places to interconnect with other nerves.
  b. "Dendritic spines" transfer information.
(3) Axon: Arises from a bulge called the **axon hillock, which leads to an initial segment. This initial segment has the smallest excitability threshold (some use these terms interchangeably).**
  a. It is likely to **depolarize at the initial segment first.**
  b. There are axonal branches at the end of the axon; each individual thread innervates a single skeletal muscle fiber. All the threads together (a branch) form a motor unit.
  c. The axon's geometry is unusual for a single cell.
    - Dimensions: **May be centimeters long or meters long but microscopically thin.**
    - Example: If you wiggle your toe, the nerve cell bodies are in your spinal cord!

iii. **Speed of conduction:** Since the axons are so long, the speed of conduction is very important. Conductions can be sped up by putting a fatlike substance known as **myelin around the neurons.** The points at which the spaces of a neuron are interrupted are known as **Nodes of Ranvier.** When an action potential is propagated down a myelinated neuron, it goes from node to node to node, etc., to the effector or integrating circuit.
  (1) Nonmyelinated: In a nonmyelinated neuron, the action potential travels at half a meter per second: **"cable conduction."**
  (2) Myelinated: In a myelinated neuron, the action potential travels at 125 to 150 meters per second: **"saltatory conduction."**
    a. Velocity is proportional to the square root of the diameter.
    b. There are 100 billion neurons in the brain alone.

iv. **Pathologies**
  (1) Multiple sclerosis: A demyelinating disease where the immune system attacks myelin or oligodendria. The myelin is replaced with scar tissue and cannot propagate action potentials correctly. A **CNS disease; progressive and often fatal.**
  -**Regeneration:** If you cut a neuron in half, the Schwann cells develop a regenerative tube and form a tunnel in which the neuron can grow back. In the PNS, the Schwann cells release chemicals (nerve growth factors) to guide the neuron back to where it is supposed to be. Oligodendria do not. In the CNS, regeneration does not readily occur, presumably because of the presence of nerve-growth-inhibiting factors. There is ongoing research to discover the right "blend" of growth factors that would make neuronal regeneration a possibility in the CNS.
  (2) Guillain-Barré Syndrome (Landry's paralysis): A disease of the **PNS** where neurons become demyelinated, usually after viral disease. Commonly **plateaus, and often there is some recovery.**

v. **Structural types**
  (1) Multipolar
    a. These have multiple dendrites and a single axon.
    b. They are found **within integrating circuits and in efferent pathways.**
    c. You typically find this kind of neuron.

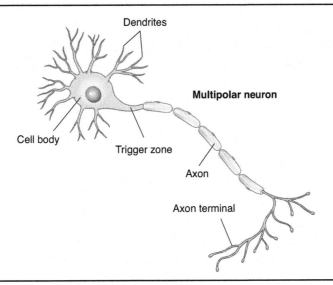

Dendrites

Multipolar neuron

Cell body

Trigger zone

Axon

Axon terminal

FIGURE 4.5

(2) **Bipolar**
   a. These have a single axon and dendrite.
   b. These are not common but are found in the olfactory mucosa.

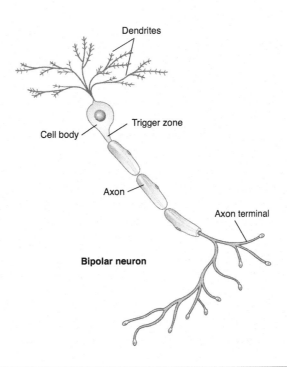

Dendrites

Trigger zone

Cell body

Axon

Axon terminal

Bipolar neuron

FIGURE 4.6

(3) Pseudounipolar
  a. These have one process, the axon.
  b. The nerve cell body sits to one side.
  c. These are **found in afferent pathways and are not uncommon.**

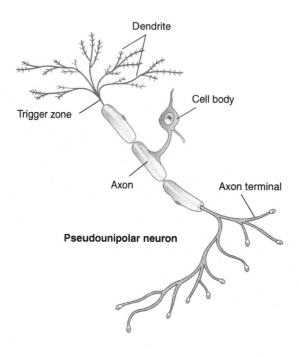

FIGURE 4.7

c. **Electrical synapses (gap junctions)**
  i. Synapse: The location where the neuron comes in contact with another excitable tissue.
  ii. Specialized synapses: The presynaptic membrane and the postsynaptic membrane are ten times closer together than normal synapse.
      (1) They essentially touch each other.
      (2) Proteins on each of the two membranes (presynaptic and postsynaptic) project outward toward each other, forming a gap junction.
      (3) There is no need for a neurotransmitter since the action potential is propagated directly across synapse.
      (4) This type of synapse is not uncommon.
d. **Chemical synapses:** The presynaptic structure releases neurotransmitter, which causes the action potential in the postsynaptic structure.
  i. **Anatomical classification**
      (1) Neuro-neuronal: The presynaptic and the postsynaptic structures are neurons.
      (2) Neuromuscular: The presynaptic structure is a neuron and the post synaptic structure is a muscle.
      (3) Neuroglandular: The presynaptic structure is a neuron and the postsynaptic structure is a gland.

ii. **Response in postsynaptic structure (ionic):** Communicate excitatory and inhibitory messages but are still transmitted by neurotransmitters.
   (1) Excitatory synapses

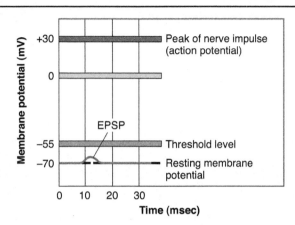

FIGURE 4.8

   a. In excitatory synapses the neurotransmitter **increases the Na$^+$ permeability** of the postsynaptic structure and causes a **hypopolarization in the resting membrane potential**.
   b. This is called **an EPSP** (Excitatory Post Synaptic Potential) when no action potential is produced.
   c. The membrane becomes less polar (**hypopolarized**) and is thus more excitable (**hyperexcitable**).
   d. If Na$^+$ permeability increases enough, then you will eventually reach threshold and get an action potential.
   (2) Inhibitory synapses

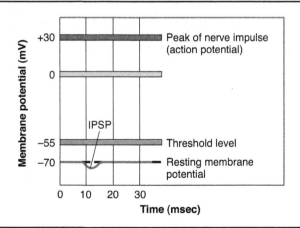

FIGURE 4.9

a. When the neurotransmitter binds to the postsynaptic membrane, **K⁺ or Cl⁻ permeability increases**. It is easier for K^+ or Cl^- to diffuse through the membrane now.

b. This causes an **IPSP** (Inhibitory Post Synaptic Potential), and the membrane is now **hyperpolarized**.

c. It is also **hypoexcitable**.

iii. **Principle of convergence and divergence.**

(1) **Divergence**

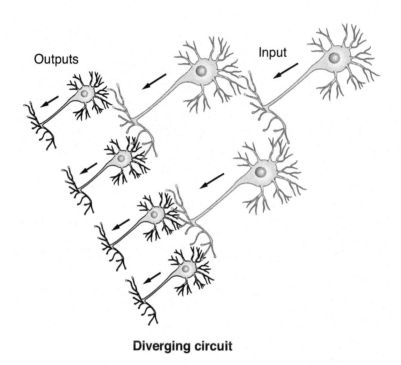

Diverging circuit

FIGURE 4.10

a. The presynaptic structure has multiple postsynaptic structures.

(2) Convergence

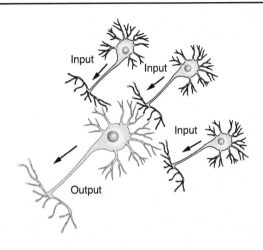

Converging circuit

FIGURE 4.11

    a. A single postsynaptic structure has multiple presynaptic structures.

(3) A single neuron could simultaneously be a presynaptic structure and postsynaptic structure. It also may be diverging to more than one structure and being converged upon by multiple presynaptic structures.

(4) Example:

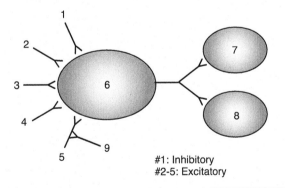

#1: Inhibitory
#2-5: Excitatory

FIGURE 4.12

    a. One neuron (6) is the postsynaptic structure for 1—5 neurons (convergence); 6 diverges to 7 and 8; 6 is a presynaptic structure and a postsynaptic structure.

    b. If the excitatory threshold of 6 is 20 millivolts and neurons 1–5 release neurotransmitter, that changes the resting membrane potential of 6 by 5 millivolts.

    c. Notice that 2–5 are excitatory, whereas 1 is inhibitory.

     d. **If you stimulate 3:**

     e. **If you stimulate 1 and 5:**

     f. **If you stimulate 1–5:**

  iv. **Implications**
- **(1) Spatial summation**: Multiple neurons stimulated at the same time: "summating in space"
- **(2) Temporal summation:** Stimulations (one after another) *from one neuron:* "summating in time"
- **(3) Presynaptic modulation:** Suppose stimulating 5 causes an EPSP of 5 mV.
  - a. **Presynaptic facilitation:** Stimulating 9 and 5 raises the voltage of #5 from 5 to 8 millivolts.
  - b. **Presynaptic inhibition:** Stimulating 9 and 5 lowers the voltage of #5 from 5 to 2 millivolts.
    It is thought that these work by altering the amount of Ca^{++} that comes into neuron 5.
  v. **Co-localization:** The idea that one neuron may release more than one neurotransmitter.
  vi. **Retrograde Messengers:** Postsynaptic structures may be releasing messengers that serve to shut off presynaptic messenger.

MID-CHAPTER 4 QUESTIONS

1. _____ Moving your little finger is what kind of function?
   A. Somatic afferent
   B. Somatic efferent
   C. Visceral afferent
   D. Visceral efferent

2. _____ Wiggling your little toe is what kind of function?
   A. A visceral efferent function carried in a cranial nerve
   B. A visceral afferent function carried in a cranial nerve
   C. A somatic efferent function carried in a spinal nerve
   D. A somatic efferent function carried in a cranial nerve
   E. A somatic afferent function carried in a spinal nerve

3. _____ The number of neurons in the afferent limb of a visceral reflex is (A) greater than, (B) less than, or (C) the same as the number of neurons in the efferent limb of a somatic reflex

4. _____ The velocity of conduction of an action potential in salutatory conduction is (A) greater than, (B) less than, or (C) the same as the velocity of conduction in a nonmyelinated fiber.

5. _____ Schwann cells
   A. Can be found in the spinal cord
   B. Can be found in neurons exhibiting salutatory conduction
   C. Are "attacked" in those suffering from multiple sclerosis
   D. All of these
   E. Two of these

6. _____ Suppose neuron A is being converged upon by four other excitatory neurons (neurons 1–4), each of which alters the resting membrane potential of neuron A by 5 mV. Also, suppose that neuron A is a presynaptic structure to neuron X and alters its (neuron X's) RMP by 10 mV by increasing its (neuron X's) permeability to potassium. The excitability threshold of both neuron A and neuron X is 20 mV. If you stimulate neurons 1 and 2, what will you observe in neuron X?
   A. An EPSP
   B. An IPSP
   C. An action potential
   D. The RMP
   E. Presynaptic inhibition

7. _____ Presynaptic facilitation is thought to occur
   A. By increasing the sodium permeability of the presynaptic structure
   B. By increasing the potassium permeability of the presynaptic structure
   C. By increasing the calcium permeability of the presynaptic structure
   D. By recruiting more motor units

IV. **Anatomical Overview of Nervous System: The Brain**
   a. Anatomical Subdivisions
      i. **Forebrain**
         (1) Cerebrum
            - Basal ganglia in white matter
            - Cerebral cortex is gray matter—has limbic system
         (2) Thalamus—has limbic system, has basal ganglia
         (3) Hypothalamus—has limbic system
      ii. **Midbrain**—has basal ganglia
         (1) Corpora quadrigemina
            (reticular formation)
         (2) Cerebral peduncles
      iii. **Hindbrain**
         (1) Cerebellum
         (2) Pons
         (3) Medulla
   b. **Brainstem = midbrain, pons, & medulla**

V. **The Forebrain**
   a. **Cerebrum**: The cerebrum is the largest and most conspicuous feature of the brain.

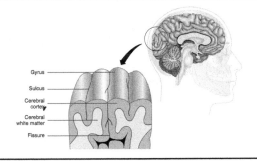

Gyrus
Sulcus
Cerebral cortex
Cerebral white matter
Fissure

FIGURE 4.13

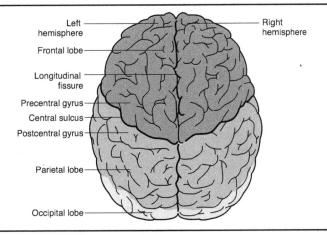

Left hemisphere
Right hemisphere
Frontal lobe
Longitudinal fissure
Precentral gyrus
Central sulcus
Postcentral gyrus
Parietal lobe
Occipital lobe

FIGURE 4.14

i. General anatomy
    (1) Outer grey matter: composed of densely packed nerve bodies and glial cells
        a. Grey matter develops faster than white matter.
        b. Grey matter folds up on itself and looks like a topographic map.
        c. Gyri or gyrus: ridges
        d. Sulci or sulcus: indentions
    (2) Inner white matter: composed of bundles of myelinated axons
    (3) There are two cerebral hemispheres: left and right.
    (4) The grey matter of each cerebral hemisphere is often further anatomically subdivided into four regions, or lobes, named after the cranial bones that cover them: the frontal lobe, parietal lobe, temporal lobe, and occipital lobe.
    (5) The outer grey matter can also be subdivided into three functional areas:
        a. Sensory area perceives sensory input.
        b. Motor area initiates skeletal muscle activity.
        c. Association area integrates information from sensory and motor areas.

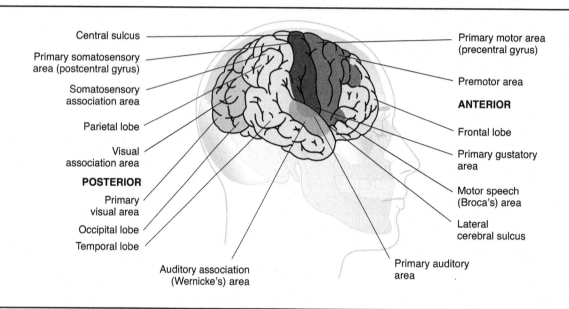

FIGURE 4.15

ii. Sensory functions
    (1) Somatosensory cortex
        i. Parietal lobe
        ii. Skin, skeletal muscle
      a. Exteroreceptive (touch, pain, temperature) and proprioreceptive (body position)
      b. Contralateral pathways cross on the way to the somatosensory cortex. Therefore, right side of brain receives information from left side of body.
      c. Homunculus

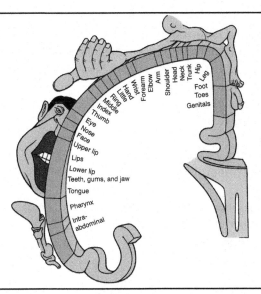

FIGURE 4.16

(2) Visual cortex (occipital lobe)
(3) Auditory cortex (temporal lobe)
(4) Olfactory cortex (small portion of temporal lobe)
(5) Gustatory cortex
iii. Motor functions
   (1) Primary motor area
      a. Found in the precentral portion of the frontal lobe
      b. Contralateral
      c. Homunculus
      d. Also receives input from the cerebellum and the basal ganglia

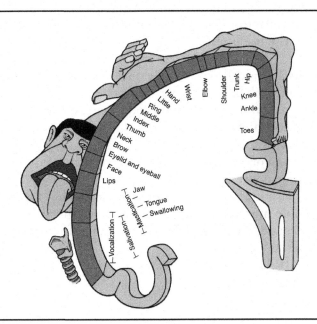

FIGURE 4.17

        (2) Language area
           a. Translates thoughts into speech
           b. Found in part of the frontal lobe called Broca's area
    iv. Association areas integrate information from both sensory and motor pathways
        (1) Somatosensory association area
           a. Stores memories of past sensory experiences
           b. Found in the parietal lobe
        (2) Wernicke's area
           a. Translates what is heard as speech into thoughts
           b. Found in the temporal lobe
        (3) Gnostic area
           a. Integrates information from several sources to create common thought
           b. Found in the parietal lobe
        (4) Prefrontal association area
           a. Initiates thought
           b. Cognitive area of the brain
    v. **Left and right side differences**
        (1) Cortical areas are equally divided between the right and left hemispheres except that **language is primarily in the left.**
        (2) **Left has the fine motor control; most people are right-handed.**
        (3) **Left side excels** in logic, analyzing, verbal tasks, and sciences.
        (4) **Right side excels** in nonlanguage skills such as art, music, and spatial skills.
    vi. **Fiber types**
        (1) **Association:** runs within the same hemisphere
        (2) **Commissural:** runs from one cerebral hemisphere to the other cerebral hemisphere
           a. **Corpus colossum:** bundle of commissural fibers
        (3) **Projection:** Fibers either carry information out of the cerebrum to other parts of the CNS or receive information from other parts of the CNS.
  b. **Basal Ganglia**
    i. **Places** (putamen, globis pallidus, substantia nigra [which is affected in Parkinson's patients]): **mostly in white matter of cerebrum, but also midbrain and thalamus; "substantia nigra" is within the midbrain.**
    ii. **Functions**
        (1) **Inhibits muscle tone.** Proper muscle tone is a balance of excitatory and inhibitory input.
        (2) Usually provides inhibitory input
        (3) **Selects purposeful muscle activity** and suppresses others
        (4) The result is **coordinated and sustained muscle contractions.**
    iii. **Parkinson's disease:** uncoordinated muscle activities such as tremors and shakes associated with the malfunction of basal ganglia in substantia nigra of midbrain (and the lack of a neurotransmitter called dopamine)
        (1) The neurotransmitter, dopamine, miscommunicates between basal ganglia, causing **uncontrollable muscles** and uncoordinated movement.
        (2) Treatments include levodopa (**DOPAR),** which is **the precursor to dopamine** and thus enhances its synthesis.
        (3) The drug (**MIRAPEX)** is similar to dopamine and **stimulates dopamine receptors** to turn on (but it is not dopamine).

c. **Thalamus**: serves as a relay station—has limbic system
   i. Lots of synapses occur here. It performs much **preliminary processing of sensory input** on its way to the parietal cortex.
   ii. It **screens out insignificant signals** and routes the more important ones to the sensory cortex.
   iii. It does not distinguish things very carefully but only processes signals from good and bad—**"crude awareness"**
d. **Hypothalamus**—has limbic system.
   i. **Temperature regulation:** maintains body temperature at 98.6 degrees Fahrenheit (37 degrees Centigrade)
      (1) A very complex system
      (2) Recognition neurons are sensitive to the temperature of the fluid bathing them.
      (3) The neurons begin a cascade to decrease or increase the temperature of the body (sweating).
   ii. **Food**: Within the hypothalamus are two centers.
      (1) One is the eating center, which causes us to be hungry.
      (2) The other is the satiety center, which causes us to become satisfied after eating.
      (3) Example: In an experiment on rats, scientists destroyed their satiety centers, and the rats never stopped eating.
         a. The scientists did the opposite, destroying the eating centers, and the rats became anorexic.
   iii. **Water:** Osmoreceptors monitor the saltiness of our blood.
      (1) If blood is too salty, then the osmoreceptors are stimulated and cause us to become thirsty. They also cause the pituitary gland to release ADH (antidiuretic hormone), which decreases urine volume.
   iv. **Sleep:** Biological clocks are influenced by day or night.
      (1) Our clocks are tuned to time.
      (2) We have cycles of activity and rest according to our biological clocks.
   v. **Hormone and releasing factors:** the **pituitary gland** is in close relationship with the hypothalamus.
      (1) Hormones of posterior pituitary gland are made in the hypothalamus. The release of hormones from the posterior and anterior pituitary gland is **governed by the hypothalamus.**
   vi. Activates the sympathetic nervous system.

VI. The Midbrain
   a. The "roof": four bumps, called corpora quadrigemina
      i. **Superior colliculi**: two of those bumps
         (1) Visual reflexes
         (2) Example: when you see movement and turn your head in response
      ii. **Inferior colliculi**: the other two bumps
         (1) Auditory reflexes
         (2) Example: when you turn your head in response to sound
   b. **The Midportion (reticular formation)**
      - Modifies sensory input
      - Arouses the cortex
   c. **The Base (cerebral peduncles):** massive nerve tracts that communicate with the cerebrum
      - Carry information to and from the cerebrum

VII. **The Hindbrain**
a. **Cerebellum**: shaped like butterfly wings; central portion is the **vermis,** and the "wings" are each hemisphere.
   i. **Vestibulocerebellum (houses "flocculondular nucleus"):** maintains balance and orientation in space
      (1) Motion sickness is a dysfunction of the flocculondular nucleus.
   ii. **Spinocerebellum:** regulates muscle movement.
      (1) Is like middle management.
      (2) Comparisons are made between what the intended muscle movement is and what is actually occurring.
   iii. **Cerebrocerebellum:** plans and initiates muscle movements, works with cerebrum.
b. **Pons**
   i. **Tegmentum:** role of conduction pathway, like the cerebral peduncles
   ii. Has **nuclei associated with respiration** and cardiovascular function
   iii. Has some **reticular formation**
c. **Medulla**
   i. **Vital functions:** acts as integrating circuit to vital functions such as respiration, heart rate (HR), and blood pressure (BP); swallowing; coughing; vomiting (**"nucleus of the tractus solitarius"/"nucleus tractus solitarius"**)
   ii. Decussation of the pyramids
   iii. **Nucleus gracilis** and **nucleus cuneatus**: important sensory pathways
   iv. **Reticular formation**

VIII. **The Limbic System**
a. Structure
   i. Not a separate structure of the brain but **a network in the brain**
   ii. There are a number of structures associated together in the forebrain and involved in a particular function of the limbic system (such as the **cerebral cortex, thalamus, and hypothalamus).**
   iii. A network of neurons (limbic) that control emotions, behavioral patterns, etc.
   iv. In an experiment, scientists stimulated the limbic systems of animals, which caused complex, bizarre behavior (such as anger and rage in normally docile animals and vice versa).
   v. A study was done by **Olds and Milner** that identified the pain and pleasure centers in our brains.
      (1) They put stimulating electrodes in different parts of the limbic system of three groups of rats. The rats determined the amount of stimuli given to them by pushing a button.
         a. **Group 1** had electrodes in their limbic system and were used as a control group. These rats hit the button 15–20 times an hour out of curiosity, but nothing happened to them; they were not stimulated.
         b. **Group 2** had electrodes in their pain centers. These rats hit the button only one time and learned that it stimulated pain.
         c. **Group 3** had electrodes in their pleasure centers. These rats hit the button 5,000–7,000 times an hour. They even stopped doing other activities (such as eating) and starved themselves.
b. **Mechanisms Responsible**
   i. The mechanisms responsible for behavior are unknown, but the hormones norepinephrine, dopamine, and serotonin are all implicated.
   ii. By pharmacologically altering the levels of neurotransmitters, diseases such as schizophrenia and depression or other emotional diseases can be treated.

iii. Pathologies
- **(1) Schizophrenia:** due to an excess of dopamine in the limbic system
  - a. Drugs used to treat this disease interfere with dopamine receptors.
- **(2) Depression:** due to too little norepinephrine or serotonin (or both) in the limbic system.
  - a. **Prozac blocks the deactivation of serotonin** (it is an SSRI—selective serotonin reuptake inhibitor).
- **(3)** There is great potential for drugs that could be developed to treat these kinds of diseases.

IX. The Reticular Formation
a. Structure
i. Runs through the entire brain stem, thalamus, and cerebrum.
b. Function
i. **Modifies sensory input** and filters extraneous information.
- **(1)** Does not bring all the information to the level of consciousness.

ii. **Arouses the cortex** and brings it to level of consciousness.

iii. Example: Electrodes hooked up to your reticular formation and to your cortex would record alpha rhythm in your brain while you sleep. If someone aroused you, then the alpha rhythm would change to beta in both locations. **It would first be seen in your reticular formation, meaning that your reticular formation would arouse your cortex.**

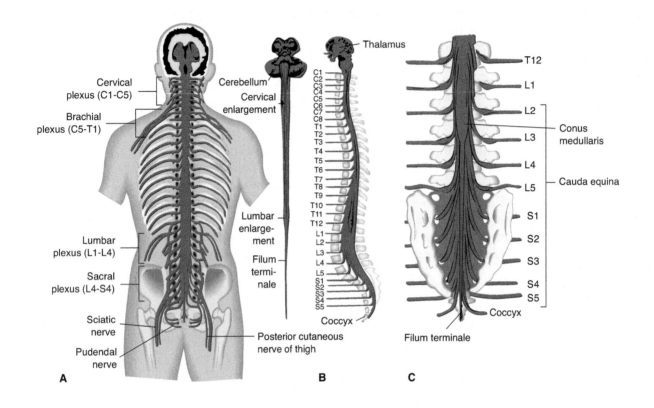

A

B

C

Labels in figure:
- Cervical plexus (C1-C5)
- Brachial plexus (C5-T1)
- Lumbar plexus (L1-L4)
- Sacral plexus (L4-S4)
- Sciatic nerve
- Pudendal nerve
- Cerebellum
- Cervical enlargement
- Lumbar enlargement
- Filum terminale
- Posterior cutaneous nerve of thigh
- Coccyx
- Thalamus
- Conus medullaris
- Cauda equina
- Filum terminale

FIGURE 4.18

a. Gross anatomy: The spinal cord is continuous from the medulla and extends to the second lumbar (or L-1/L-2 IV disc).
   i. It is lodged **within the spinal canal.**
   ii. It is only as thick as a pencil (1–2 cm).
   iii. Coming off the spinal cord are pairs of spinal nerves that come out between vertebrae.

(1) Each is named according to the root of entry or exit:
  a. **8 pairs of cervical** spinal nerves
  b. **12 pairs of thoracic** spinal nerves
  c. **5 pairs of lumbar** spinal nerves
  d. **5 pairs of sacral** spinal nerves
  e. **1 pair of coccygeal** spinal nerves
  f. **31 pairs of spinal nerves in total**
     - The portion below L-2 is known as the **cauda equina:** Nerves run caudally ("toward tail") before they run laterally.
     - **Spinal nerves have both visceral and somatic functions.**
  b. Microscopic anatomy

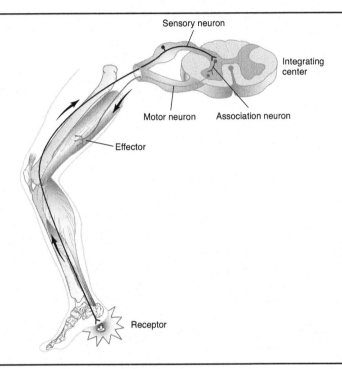

FIGURE 4.19

  i. This is a cross-section of the spinal cord.
     (1) The "butterfly" part is the **gray matter,** and it consists of **nerve cell bodies and dendrites of neurons.**
     (2) The **white matter is made up of myelinated axons.**
     (3) The pointed part of the "butterfly" is called the **dorsal horn** of the gray matter and is **posterior.**
        a. **The dorsal horn is associated with afferent functions.**
     (4) The bottom horn is called the **ventral horn,** and it is **anterior.**
        a. **The ventral horn is associated with motor functions (efferent).**
     (5) In the above picture we see the **pseudounipolar neuron** that is in the afferent pathway (the cell body is the **"dorsal root ganglion"**).
     (6) A **multipolar neuron** is in the integrating circuit and the efferent pathway (as shown in the picture above).

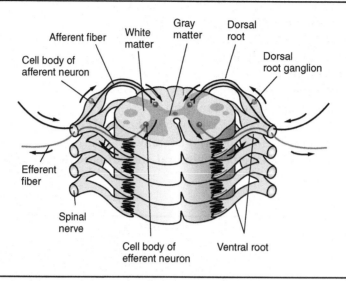

Labels in figure:
Afferent fiber
Cell body of afferent neuron
White matter
Gray matter
Dorsal root
Dorsal root ganglion
Efferent fiber
Spinal nerve
Cell body of efferent neuron
Ventral root

FIGURE 4.20

ii. In the CNS, neurons carrying information up the spinal cord are parts of ascending pathways, and neurons carrying information down the spinal cord are parts of descending pathways.

   (1) The exact point where an afferent neuron becomes ascending and where a descending neuron becomes motor is ambiguous.

iii. In the middle of the gray matter is a hole, **the central canal.**

   (1) The central canal is continuous with the ventricles (specifically, **ventricle IV**) in the brain (via the **foramen of Magendie).**

   (2) The white matter is organized into various tracts (some ascending and some descending).

      a. There are certain tracts for various functions.

      b. Example: Some are associated with certain types of sensations.

   (3) The pathways are always in the same place throughout the spinal cord.

   (4) What is the consequence of a complete severing of the spinal cord?

   (5) What is the consequence of an incomplete severing of the spinal cord?

   (6) What is the consequence of the spinal cord being severed in the lumbar region versus the cervical region?

   (7) How can some quadriplegics move their arms?

XI. **Protection for the CNS**

   a. Bone

      i. The brain and the spinal cord are housed within a bony vault.

         (1) The brain is housed in the **cranial cavity.**

         (2) The spinal cord is housed in the spinal canal, within the **vertebral column.**

b. The **meninges**
   i. **Meninges**: collective term referring to the three membranes that protect the CNS
   ii. **Dura mater** ("tough mother"): a tough, inelastic covering that is continuous with the **periosteum** in the cranial cavity (no epidural space)
      (1) In the spinal cord, there is a space between the periosteum and the dura mater called the **epidural space.**
         a. Injecting anesthetic into the epidural space affects the body according to the area near the spinal cord and spinal nerves at which it is injected.
      (2) In the brain, the dura mater is against the periosteum; they are "stuck" to each other.
         a. At some places the two layers of dura mater "come apart," forming dural sinuses (Figure 4.21). These dural sinuses play a role in circulating CSF. **The dural sinuses are drained venously into the bloodstream.** Thus, we will see later that CSF is taken up from the **subarachnoid space** by the **arachnoid villi**, where they empty the CSF into the dural sinuses for **return to the venous system.**
      (3) There is an extension of the dura mater called the **falx cerebri.** This extension separates the cerebrum into left and right hemispheres.
   iii. **Arachnoid mater:** a filamentous (web-like) membrane that separates the dura mater from the pia mater
      (1) Arachnoid mater has **arachnoid villi (projections)** that project into the dural sinuses. As mentioned, these play a role in draining CSF back to the venous system.
   iv. **Pia mater** ("gentle mother"): This is highly vascularized and sticks to the brain and the spinal cord (**directly to the nervous tissue**).
      (1) In some places the pia mater "digs" deeply into the brain to bring blood into close association with the brain.
      (2) This brings the blood into close contact with specific glial cells called **ependymal cells,** which produce the CSF from an area in the ventricle called the **choroid plexus.**
   v. Between the arachnoid mater and the pia mater is a space called the **subarachnoid space (present in both brain and spinal cord).**

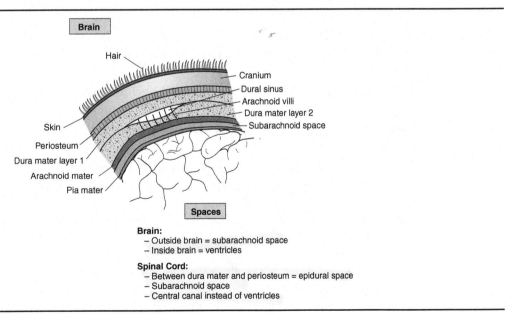

FIGURE 4.21

c. **Cerebral Spinal Fluid**
   i. CSF is located in the ventricles, subarachnoid space, and central canal.

Ventricles inside the brain

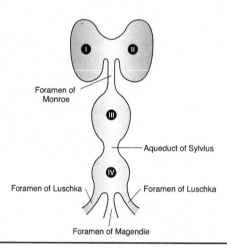

FIGURE 4.22

   (1) Ventricles are the (hollow, in a way) cavities inside the brain.
   (2) There are four ventricles, and they are connected to each other.
   (3) **Foramen of Monroe** connects I and II to III.
   (4) The **Aqueduct of Sylvius** connects III to IV.
   (5) They are also connected to the central canal (hole in spinal cord) by the **Foramen of Magendie.**
   (6) The **Foramen of Luschka** connects IV to the subarachnoid space.
   ii. **Choroid plexus and CSF**
   (1) The brain can become bruised if it is hit because it **bounces off the opposite wall of the cranial vault (inertia).** Cerebrospinal fluid helps to protect the brain.

   Why do boxers have their eyes checked periodically during matches?

   How could being hit in the face damage the back of the brain??

   (2) CSF is made in gland-like structures called the **choroid plexus.**
      a. The choroid plexus is found in "various places in the ventricles" (it is truly a net-work) and **continually produces CSF.**
   (3) CSF travels through the foramina to protect the brain (via the subarachnoid space) and down the central canal.
      a. Example: The brain is like "a submarine in the CSF." The CSF is also inside the brain in ventricles. It travels down the central canal and also acts as a shock absorber in subarachnoid space.
   (4) CSF is continually being produced and is drained through the arachnoid villi, which project into the dural sinuses. CSF then is recycled to the general circulation via venous drainage.

      iii. Hydrocephaly ("water head"): Typically a congenital problem in the ventricles or ventricular circulation; e.g., I and II cannot drain into III, and so on (a foramen or duct is blocked).

          (1) The CSF cannot flow freely but it is still continually being produced in ventricles I and II!

              a. It has no way to be drained—the pressure increases, and the brain begins to expand.

              b. Sutures have not yet formed in young children, so the head becomes enlarged.

              c. Can put a shunt in to drain off excess CSF.

      iv. "Spinal tap": a procedure to collect CSF in order to look for bacteria (meningitis, etc.)

          (1) Where should we insert a needle to draw a CSF sample?

              a. Ventricles?

              b. Central canal?

              c. Subarachnoid space in brain or spinal cord?

              d. Where would you like CSF collected if a first-year resident were inserting the needle?!

XII. Blood-Brain Barrier

    a. Example: Scientists injected vital dyes (colored) into animals to see how much circulated in different areas.

      i. They could not find dyes that went to the brain or spinal cord–they had discovered **the blood-brain barrier.**

    b. The blood-brain barrier is between the circulatory system and the brain or spinal cord.

      i. It keeps pathogens from getting into the CNS.

    c. If you get a malady in the CNS, you cannot get many drugs into the CNS to treat the malady.

XIII. The Spinal Nerves

    a. Functional classification

      i. Four classes:

          (1) GSA: General Somatic Afferent

              a. All but the C-1 and the coccygeal spinal nerves

              b. Receives information from the soma

              c. Example: Burns on extremities

          (2) GSE: General Somatic Efferent: Impulses to the soma

              a. All the spinal nerves

              b. Effector in soma is skeletal muscle

              c. Example: Moving fingers to take notes

          (3) GVA: General Visceral Afferent

              a. T-1 to L-3 and S2, S3, and S4

              b. Impulses coming back from the viscera

          (4) GVE: General Visceral Efferent

              a. Same as above

              b. Effector is smooth muscle, cardiac muscle, or a glandular structure

              c. Efferent reflex: two pathways

      ii. Wait, how can a nerve have more than one function?

      iii. All spinal nerves have somatic function, and nearly all have mixed efferent and afferent functions.

(1) Nerves have different neurons: afferent and efferent.
   a. Notice the neurons and types in Figure 4.20.
(2) Although spinal nerves have mixed functions, the spinal cord has both the dorsal horn, which houses the afferent neuron entrance, and the ventral horn, which houses the efferent neuron exit.

b. Distribution: 31 pairs of spinal nerves innervate the soma (below the neck), and each spinal nerve has its own responsibilities to innervate the **dermatome (the portion of the soma innervated by a single spinal nerve)**.

   i. Different spinal nerves innervate different parts of the body. Notice that in Figure 4.23 you will see the **"29 dermatomes."** *Dorland's Medical Dictionary* defines a dermatome as an **"area of skin supplied with afferent nerve fibers by a single posterior spinal root."** This is a clinically useful chart that shows the afferent innervation of each spinal nerve. However, **all 31 spinal nerves have efferent function!**

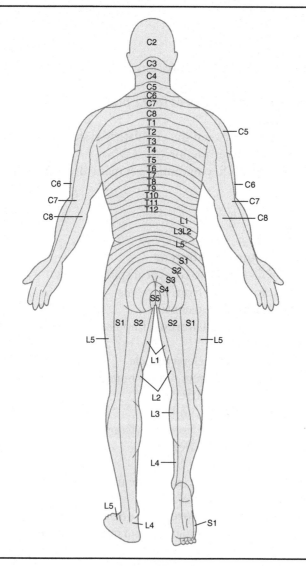

FIGURE 4.23

XIV. The Cranial Nerves
a. Functional Classifications
   i. These are distributed to structures of the head and neck except for the vagus nerve (the wanderer), which goes down to structures like the small intestine (SI).
   ii. GSA, GSE, GVE, GVA: There are also special senses associated with some of these nerves that are highly specialized.
      (1) SSA: Special Somatic Afferent, associated with hearing and seeing **(nerves 2 and 8)**
      (2) SVA: Special Visceral Afferent, associated with tasting and smelling **(nerves 1, 7, and 9)**
      (3) SVE: Special Visceral Efferent, associated with muscles of mastication (chewing) **(nerve 5)**
         a. "Visceral" because derived embryologically from visceral muscles.
b. The Cranial Nerves
   i. Olfactory (I)
      (1) Smell
   ii. Optic (II)
      (1) Vision
   iii. Oculomotor (III)
      (1) Motor
         a. Movement of eyelid and eyeball
         b. Accommodation of lens
         c. Constriction of pupil
      (2) Sensory
         a. Proprioception from eye muscles
   iv. Trochlear (IV)
      (1) Motor
         a. Movement of eyeball
      (2) Sensory
         a. Proprioception of eye muscles
   v. Trigeminal (V)
      (1) Motor
         a. Chewing
      (2) Sensory
         a. GSA from nose, forehead, scalp, teeth, jaw, cheek, etc.
   vi. Abducens (VI)
      (1) Motor
         a. Movement of eyeball
      (2) Sensory
         a. Proprioception of eye muscles
   vii. Facial (VII)
      (1) Motor
         a. Facial expressions
         b. Secretion of saliva and tears
      (2) Sensory
         a. Taste
         b. Proprioception from muscles of face and scalp
   viii. Vestibulocochlear (VIII)
      (1) Hearing and equilibrium

  ix. Glossopharyngeal (IX)
   (1) Motor
    a. Secretion of saliva
    b. Muscles of throat and larynx
   (2) Sensory
    a. Taste
    b. Regulation of ventilation and BP
    c. Proprioception
  x. Vagus (X)
   (1) Motor
    a. Smooth muscle contraction and relaxation in several visceral structures (airways, esophagus, stomach, SI)
    b. Decrease HR
    c. Secretion of digestive fluids
   (2) Sensory
    a. Sensations from visceral organs
    b. Regulation of ventilation and BP
  xi. Accessory (XI)
   (1) Motor
    a. Swallowing
    b. Movement of head
   (2) Sensory
    a. Proprioception
  xii. Hypoglossal (XII)
   (1) Motor
    a. Movement of tongue
   (2) Sensory
    a. Proprioception from tongue muscles

Let's review! What can you tell me about...

GSE –

GSA –

GVE –

GVA –

SSA –

SVA –

SVE –

If you cannot quickly rattle off many things associated with these terms, stop here and review the neurophysiology thus far. It is paramount that you master this material before moving on!

XV. Receptor/Sensory Physiology

   a. Receptor types: Design of receptor is made to respond to a particular stimulus. You cannot "hear" the light I shine in your ear!

   i. **Mechanoreceptors:** These are sensitive to mechanical stimulation.
      (1) They change shape and send impulses.
      (2) Example: The Organ of Corti (ear): hairs wiggle and start the hearing/auditory pathway.
      (3) Example: BP regulation is via stretch—mechanoreceptors.

   ii. **Thermoreceptors:** These sense changes in the temperature.
      (1) Example: Receptors in the hypothalamus, skin

   iii. **Nociceptors:** These are stimulated by damage done to the receptors.
      (1) Pain receptors
      (2) They are activated by damage around their tissue.

   iv. **Electromagnetic receptors:** These are sensitive to electromagnetic radiation.
      (1) Example: Occurs in our eyes as we perceive light. Some animals are sensitive to magnetic fields.

   v. **Chemoreceptors:** These are sensitive to the body's chemistry.
      (1) Example: Measure O_2, CO_2, H^+, Na^+

   b. **Weber-Fechner principle**

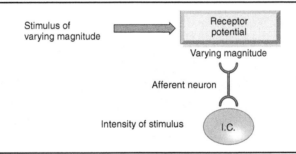

FIGURE 4.24

   i. As the magnitude of the stimulus becomes greater, the magnitude of the receptor potential becomes greater (not by a linear increase, but **by a logarithmic increase**).
      (1) In order to get a linear change in the receptor potential, you have to have a logarithmic change in the stimulus intensity.

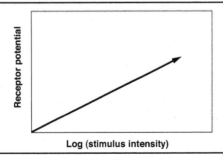

FIGURE 4.25

The relationship shown above is the "Weber-Fechner principle."

ii. Receptor potential (an electrical phenomenon that is graded, not "all or none") dictates frequency of action potentials in afferent neurons, which in turn dictates how strongly the intensity is perceived when it terminates in the brain.

  (1) The bigger the receptor potential, the more frequently you get action potentials from the afferent neuron!! This (the frequency) is not "all or none" but a graded response!

iii. There are limits to the higher end of a frequency of action potentials on afferent side. **What limits the action potentials that can be sent afferently from the receptor potential (or, ultimately, the perception of the strength of that stimulus)?**

iv. **What is the relationship between the stimulus that is applied and the stimulus that is perceived?**

v. If it were a linear function instead of a log function, we could perceive very precisely in a very narrow range! **We instead use the log function, because ...**

c. **How do different receptors detect different stimuli?** How come we see with our eyes and hear with our ears but not vice versa? How do receptors detect only one type of stimulus?

  i. This is all due to **the design of the receptor** and the different types of stimuli.

    (1) Example: The Organ of Corti: In response to sound waves, hairs move back and forth.

    (2) Example: Eyes have rods and cones that detect light stimulation.

  ii. This is all due to the design of the receptors and the fact that receptors respond electrically by developing a receptor potential (an electrical event of varying magnitudes).

    (1) **This is not like an action potential** ("all or none").

  iii. Receptor potential ultimately causes an action potential in the afferent neuron.

d. **Modality of sensation: How are we able to detect different types/modalities of sensation?**

  i. Neurons **terminate in different parts of the brain,** and the brain interprets specific signals.

    (1) Example: When you receive a blow to the head, you see "stars" (or, rather, flashes of light).

      a. Why? Because the blow mechanically stimulates the optic nerve, and an action potential is propagated to the brain.

      b. **The action potential from the optic nerve means "light" to the brain!**

    (2) **How does this explain the concept of phantom pain? What would you tell a patient experiencing phantom pain? ("Get over it—you have no legs to feel pain in!"—?)**

  ii. Stimulation receptors are *not* "all or none," but action potentials are.

e. **Intensity of sensation:** How do we experience different intensities of stimulation?
    i. The same number of neurons are stimulated each time, but the frequency of action potentials varies (still "all or none," though).
      (1) The frequency of action potentials reaching the brain determines the amount of sensation.

f. **Receptor adaptation**

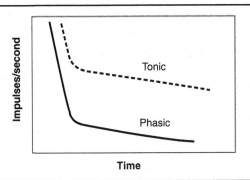

FIGURE 4.16

    i. Impulses/second means **impulses per second** in afferent neurons.
    ii. **Phasic:** These receptors, if exposed to the same stimulus constantly over time, respond rapidly at first to the stimulus, and then do not respond to it anymore.
      (1) Touch receptors adapt rapidly (such as when you first put clothes on). **This is *not* fatigue**—with a different stimulus, these will immediately respond. You may not feel your shirt until the wind blows and moves it.
      (2) The olfactory mucosa adapts rapidly in response to constant stimulation from one source.
    iii. **Tonic:** These receptors do not adapt easily.
      (1) Example: Maintain firing, do not adapt to lower firing rate

You are running down a stairway—do you want the proprioceptive feedback to be phasic or tonic?

XVI. **Ascending Pathways**
   a. **The dorsal column system**
     i. Functions
       (1) Exteroreceptive function: brings in information from outside of the body, which is integrated in the brain at the **parietal lobe**
         a. Involves **touch, especially with fine sensation**
         b. Example: Blindfold someone and give him or her objects to perceive, or stick someone with pins to test detection of the number of pins.

(2) **Proprioception:** deals with relationship of different body parts to each other, such as their orientation and rate of movement
   a. This is all **integrated in the cerebrum.** Some are conscious; some are not.
   b. There are two different parts of the brain (or integrating circuit) involved in this process.
   c. Modalities of sensation are in different parts of the brain.
   d. Pseudounipolar neurons: afferent neurons
  ii. **Anatomy (See Figure 4.27):** The picture is drawn as if a stimulus is coming in from only one side and dermatome and detected by one spinal nerve.
   (1) Usually you have more than one stimulus coming in.
   (2) Runs in the **fasciculus cuneatus** if anterior or **fasciculus gracilis** if posterior. Synapses in medulla at the corresponding nucleus (see cartoon) and then travels across the medulla.
   (3) **Synapses in the thalamus** before it travels to the sensory cortex.
   (4) **Contralateral!**
 b. **The spinothalamic system:** afferent and association neurons
   i. **Functions:** More general, such as pain, thermal sensations, crude touch (not as discriminatory or precise), tickle, itch, etc.
  ii. **Anatomy (See Figure 4.28)**
   (1) Synapses first in the grey matter of the spinal cord, then travels to the ventral posterolateral nucleus in the thalamus.
   (2) Three neurons.
   (3) Ends in the parietal lobe.
 c. **The spinocerebellar pathway:** afferent and association neurons
   i. **Functions:** informs the cerebellum of muscle performance
   (1) **Proprioceptive**
   (2) Travels from out in the dermatome to the cerebellum
   (3) May be **contralateral** or **ipsilateral**
  ii. **Anatomy (see Figure 4.28)**

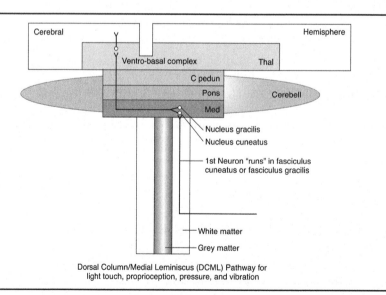

Dorsal Column/Medial Leminiscus (DCML) Pathway for light touch, proprioception, pressure, and vibration

FIGURE 4.27

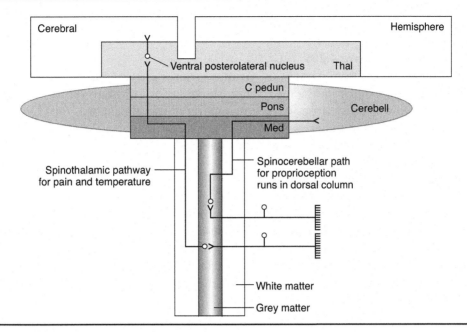

FIGURE 4.28

d. Afferent vs. ascending
   i. **Afferent neurons are found in the PNS, and ascending pathways are found in the CNS.**

XVII. Descending Pathways

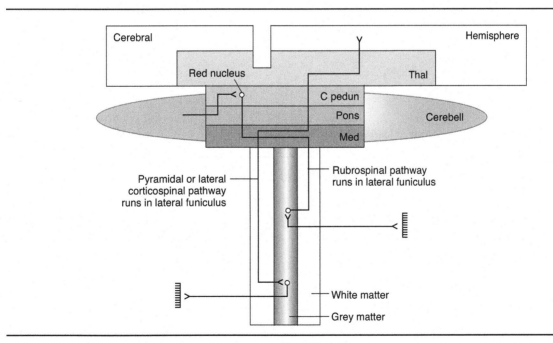

FIGURE 4.29

a. **Pyramidal ("corticospinal") pathway**
   i. **Functions:** Used in fine muscle movements
      (1) Example: Taking notes
   ii. **Anatomy (See Figure 4.29)**
      (1) Descending pathway starts in the **primary motor cortex** and comes out in the medulla.
      (2) Pathways cross over in the medulla = **"decussation of the pyramids" (landmark)**
      (3) Comes down and synapses in the grey matter in the spinal cord
      (4) A multipolar neuron here
      (5) GSE: Pathways are contralateral and cross in the medulla.
b. **Rubrospinal Pathway**
   i. **Function:** communication between the cerebellum and the red nucleus
      (1) It is the pathway's job to correct muscle movement.
      (2) The motor cortex sends info down the pyramidal pathway and also to the cerebellum. The cerebellum thus has an idea of **what should be going on.** The cerebellum will then get info from the spinocerebellar pathway as to **what is going on.**
      (3) The cerebellum now has two pieces of information. If the right movement is not occurring or certain areas are not doing what they are supposed to be doing, then this pathway corrects the muscle movement.
   ii. **Anatomy (See Figure 4.29)**
      (1) Technically the pathway <u>starts</u> in the red nucleus of midbrain (thus, it is really a two-neuron pathway!).

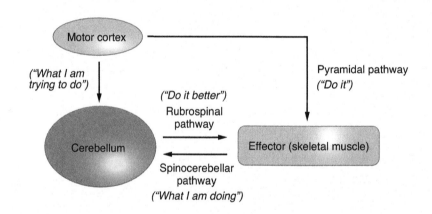

FIGURE 4.30

Figure 4.30 shows the relationship between the pyramidal pathway and the rubrospinal pathway. Note that in reality the rubrospinal pathway technically begins with the red nucleus, *not* the cerebellum.

c. **Descending vs. efferent**
   i. **Descending pathways are found in the CNS, and efferent neurons are found in the PNS.**

XVIII. Other Neurotransmitters
(ACH will be discussed more later)

Biogenic Amines
- a. **Catecholamines**
  - i. **Epinephrine**
  - ii. **Norepinephrine**
  - iii. **Dopamine**: both excitatory and inhibitory
    - (1) It is important in the **basal ganglia** and is also found in the **limbic system.**
    - (2) Has to do with the behavior of motivation and the drive or reward mechanism
    - (3) **One drug that enhances dopamine transmitters is cocaine.**
- b. **Serotonin:** both excitatory and inhibitory
  - i. This neurotransmitter is very versatile—has up to 16 different receptor types.
  - ii. It may be responsible for defect in affective disorders such as depression.
  - iii. Can be manipulated when you treat these disorders (increase serotonin levels via SSRIs or other drugs).
- c. **Histamines:** We think of these in response to allergic situations, but there are other, different types of receptors for them.

Amino Acid Derivatives
- a. **Glutamate and aspartate:** Excitatory in the brain
  - i. Found in the CNS
  - ii. Have a wide array of receptors
  - iii. Example: Angel dust (PCP) activates receptors for glutamate.
- b. **GABA:** Inhibitory neurotransmitter in the CNS (more in the brain)
  - i. Involved in the cause of anxiety
  - ii. Anti-anxiety drugs like **benzodiazepines** enhance GABA production in neurons by activating GABAnergic neurons.
  - iii. **Tetanus toxin may prevent release of GABA.**
- c. **Glycine:** Acts as an inhibitory neurotransmitter (more in the **spinal cord**) by altering Cl⁻ permeability.
  - i. Afferent pathways have ways to inhibit transmission.
  - ii. There is a lot of sensory information that may excite the membrane, but the impulses are filtered in the spinal cord and the thalamus.
  - iii. Sometimes the information is "shut off" and is not enough to cause a reaction or recognition.
  - iv. **Strychnine:** competes for glycine receptors; thus, you get way too much excitement with strychnine!
  - v. A balance between excitation and inhibition exists.

Neuropeptides
Molecules that have two or more amino acids linked together and are relatively small. These are made in the nerve cell body and released.
- a. **Substance P:** found in high concentrations in the spinal cord and the hypothalamus
  - i. Found in neurons that **transmit impulses of pain**
  - ii. **Capsaicin: depletes substance P** and thus blocks some pain.
- b. **Opioids**: endorphins, enkephalins, dynorphins; bind to same receptors as drugs such as opium and morphine. These are naturally occurring.

      i. Give a **sense of euphoria or well-being**

      ii. Example: Runner's high has been thought to be due to endorphins.

Others

    a. **Nitric oxide (NO):** a newcomer to neurotransmitters

      i. Formed from arginine and is lipid soluble, so it can diffuse in and out of cell membranes

      ii. Its response is brief; it is active for about 10 seconds.

      iii. The molecule is quite reactive.

      iv. **Relaxes endothelial walls**

      v. NO does many things: it is found in neurons of cerebrum, posterior pituitary gland, superior and inferior colliculi.

      vi. It does not always cause **vasodilation** (although it does in many vascular areas).

      vii. Responsible for **erection of the penis** (Viagra increases NO levels)

XIX. Pathologies

    a. **Neuritis:** Inflammation of nerves or neurons due to a traumatic event, pathology, drugs

      i. The pain is called **neuralgia.**

      ii. A number of nerves can become inflamed.

        (1) **Sciatica:** inflammation of sciatic nerve

        (2) **Tic douloureux:** trigeminal neuralgia

    b. **Bell's palsy:** paralysis of some or all of the facial musculature

      i. **Paralysis of the VII cranial nerve**

      ii. People *do* recover

    c. **Shingles:** caused by chickenpox virus

      i. Lifelong immunity to the chickenpox virus, but the virus is not completely destroyed and exists in a dormant state.

      ii. It travels through some dermatome back to the spinal nerve and **takes root in the dorsal root ganglion** (before the dorsal horn).

      iii. It may remain dormant for years, but if immune system is weakened (due to stress or growing older) the virus becomes active again and travels over sensory nerves and **causes excruciating pain (ulcers and blisters) on that particular dermatome.**

      iv. There is no cure.

      v. Treatment is through pain relief and steroids.

      vi. There are **times of exacerbation and times of remission.**

    d. **Cerebral palsy:** disease that **damages the motor parts of the brain**

      i. Usually damage occurs early in childhood; can be caused by oxygen deprivation during birth or by trauma to the baby's head

      ii. Causes serious motor problems

      iii. There is nothing wrong with the afferent components or integrating part; **persons with cerebral palsy are not always mentally impaired as well!**

    e. **Tumors**

      i. **Neuroma:** a tumor in the nervous system

      ii. **Glioma:** tumor of the glial cells; the most common of brain tumors; life-threatening

        (1) As it grows, it interferes with neuronal functions.

      iii. **Multiple neurofibromatosis:** an inherited disease, characterized by numerous fibrous neuromas associated with the **Schwann cells**; causes **disfiguring tumors**

f. **Cerebral vascular accident (CVA), or stroke:** the most common brain disorder and the **third leading cause of death**
   i. This is when a blood vessel to the brain has a blockage or bursts.
   ii. Blood clot may break off and block blood flow (**embolytic event**).
   iii. Strokes are **classified by severity** (transient ischemic attack, or TIA, is least severe).
g. **Seizure disorder: second most common disorder of the nervous system**; characterized by sudden, uncontrolled bursts of abnormal neuronal activity and abnormal behavior
   i. Can be mild or severe
   ii. Jerky, involuntary muscle movements
   iii. Can cause unconsciousness
   iv. **Epilepsy:** constant, frequent seizures
   v. May be motor, sensory, or both
   vi. "Spots" are the areas in the brain of spontaneous activity.
   vii. Treat with anticonvulsive drugs, which block the afferent transmissions to the brain. **You must get the dosage in a narrow range; otherwise, you may "stupefy" them (really depress neural activity).**
h. **Meningitis:** inflammation of the meninges; can be of **bacterial origin**
   i. Causes fever, stiff necks, and seizures
   ii. ~20% mortality rate
   iii. Treated with antibiotics
i. **Encephalitis:** inflammation of the brain tissue itself
   i. **Virally mediated** type is the most common.
   ii. Has an intermediate host such as an arthropod (an insect, e.g., a tick).
   iii. More severe and with a **higher mortality than meningitis**
j. **Reye's syndrome: a type of encephalitis (rare)**
   i. Also causes dysfunctions of the liver
   ii. Young children would develop Reye's syndrome just a few weeks after recovering from a viral disease; syndrome was found to be correlated with their taking aspirin. **Look at your aspirin bottle: it might mention Reye's. Now, of course, we give children acetaminophen or ibuprofen instead of aspirin.**
k. **Alzheimer's disease:** affects 4 million Americans and is age related
   i. Accounts for two-thirds of cases of age-related dementia
   ii. Short-term memory is affected first, then long-term memory
   iii. Progressively debilitating—4 to 12 years life expectancy after onset
   iv. Can be definitively diagnosed only at autopsy
   v. Affects 10% of people 60 to 65 years old
   vi. Affects 35% to 45% of people >85 years old

SUGGESTED READINGS

1. Guyton, A.C., and Hall, J.E. 2011. Chapters 45–48 and 54–58 in *Textbook of Medical Physiology*, 12th ed. Baltimore: Saunders.
2. Koeppen, B.M. and Stanton, B.A. 2008. Chapters 4-6 and 9-11 in *Berne & Levy Physiology*, 6th ed. St. Louis: Mosby.

CHAPTER 4 QUESTIONS

1. _____ Associated with the hypothalamus
   a. Substantia nigra
   b. Suprachiasmatic nucleus
   c. Temperature regulation
   d. All of these
   e. Two of these

2. _____ The flocculondular nucleus
   a. Is found in the vestibulo-cerebellum
   b. Is associated with the rubrospinal pathway
   c. Is found in the cerebrum
   d. All of these
   e. Two of these

3. _____ Associated with the PNS
   a. Schwann cells
   b. Cauda equina
   c. Ganglia
   d. All of these
   e. Two of these

4. _____ The IV cranial nerve
   a. Has something to do with the eye
   b. Has an SSA function
   c. Has an SVE function
   d. Two of these
   e. None of these

5. _____ Has an SVA function
   a. Cranial nerve IX
   b. Facial nerve
   c. Hypoglossal nerve
   d. All of these
   e. Two of these

6. _____ The spinal cord
   a. Would have pseudounipolar neurons associated with the dorsal horn
   b. Would have ventricles in the grey matter
   c. Contains some of the choroid plexus
   d. All of these
   e. Two of these

7. _____ Dural sinuses
   a. Are found between the dura mater and the periosteum
   b. Connect the subarachnoid space to the IV ventricle
   c. Contain arachnoid villi
   d. All of these
   e. Two of these

8. _____ Associated only with somatic function
   a. Dermatome
   b. Pyramidal pathway
   c. Cervical spinal nerves
   d. All of these
   e. Two of these

9. _____ Neurotransmitter affected by the benzodiazepines
   a. GABA
   b. Glycine
   c. Serotonin
   d. Glutamate
   e. NO

For questions 10–13, choose one of the following:
   A. Increases or is greater than
   B. Decreases or is less than
   C. Has no effect or is equal to

10. _____ The number of spinal nerves with a somatic function, as compared to the number of spinal nerves containing multipolar neurons

11. _____ The number of neurons in the pyramidal pathway, as compared to the number of pairs of lumbar spinal nerves with a visceral function

12. _____ The likelihood of the IV cranial nerve having an afferent function, as compared to the likelihood of the VIII cranial nerve having an SSA function

13. _____ The number of neurons in the dorsal column system, as compared to the number of neurons in the spinal thalamic pathway

CHAPTER 5

Neurophysiology II: Autonomic Nervous System

I. **Introduction to Sympathetic and Parasympathetic**
   a. Sympathetic and parasympathetic systems work toward the same goal: to maintain homeostasis within the viscera.
   b. Autonomic nervous system uses the same anatomical structures as the somatic nervous system.
   c. There are several differences between the sympathetic and parasympathetic nervous systems.
   d. **Nerves** are made up of many neurons' axons.

II. **Differences in Distribution**
   a. PNS neurons are like streets going to and coming from the CNS. Sympathetic and parasympathetic pathways never use the same "street."
   b. **Parasympathetic**
      i. **Cranial Nerves:**
         III: Functions of the eye
         VII: Functions of salivation and tear glands
         IX: Inferior salivation and lacrimal glands
         X: Vagus
      ii. **Sacral spinal nerves** (S2, S3, S4)
         (1) Pelvic viscera: genitalia, bladder, colon, rectum
   c. **Sympathetic**
      i. Sympathetic functions are general; not discrete.
      ii. **Spinal nerves (T-1 to L-3)**
         (1) 31 parts of body have sympathetic efferent function, yet those parts are sympathetically innervated from only 15 spinal nerves.
      iii. **Cranial nerves do not have sympathetic function.**

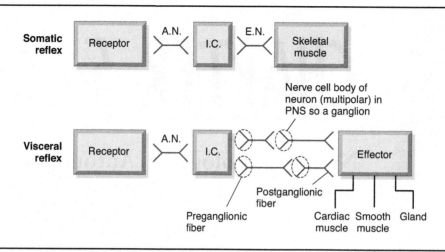

FIGURE 5.1

III. Difference in Length of Fibers and Location in Efferent Pathway of Ganglia

a. **Sympathetic**
   i. **Paravertebral:** close to the vertebrae; 24 of these
      (1) Cervical: 3—**Superior cervical, middle cervical, and inferior cervical**; these service visceral structures and all of the cervical dermatomes.
      (2) Thoracic: 12
      (3) Lumbar: 5
      (4) Sacral: 4
      (5) GVE: sympathetic are only from T-1 to L-3 but they go up or down and **diverge to ganglia.**
      (6) There are **24 ganglia for 31 postganglionic fibers servicing visceral structures and 31 dermatomes (from 15 spinal nerves).**
   ii. **Collateral (prevertebral):** Usually these are located on the dorsal aorta or its branches, farther from the spinal cord.
      (1) **Celiac**—to stomach, spleen, pancreas.
      (2) **Superior mesenteric**—to intestines.
      (3) **Inferior mesenteric**—to rectum, bladder, reproductive tract.
   iii. Preganglionic fibers exit from thoracic or lumbar area, but **they diverge to other areas.**
   iv. By diverging, preganglionic fibers come out of only 15 spinal nerves to all 31 dermatomes.
   v. There are only 27 ganglia and 31 body segments. Pathways going to all the segments arise from either paravertebral or collateral ganglia.
   vi. The big ganglia act as distribution points (these are fused ganglion of many postganglionic neurons).
   vii. The preganglionic fiber is short and the postganglionic fiber is relatively long.
   viii. Chain ganglia: paravertebral ganglia.
   ix. **Sympathetic tends to be a generalist** in its visceral functions, and **parasympathetic tends to be discrete.**

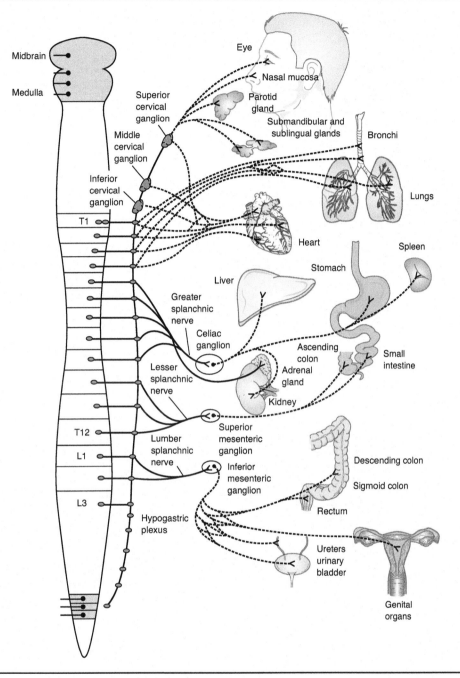

FIGURE 5.2

Sympathetic Innervation

b. **Parasympathetic (terminal):** Nerve cell body is located within the organ being innervated.
  i.   Distributed from the **cranial nerves or the sacral spinal nerves.**
  ii. The preganglionic fibers are relatively long and the postganglionic fibers are relatively short.
c. Some ganglia are fused.

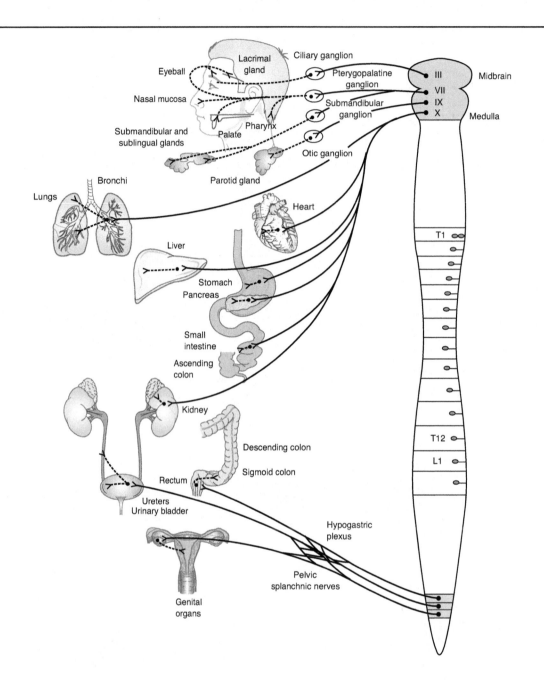

FIGURE 5.3

Parasympathetic Innervation

IV. Differences in Neurotransmitters and Receptors: Cholinergic (Acetylcholine Secretors)
    a. **Synthesis of Neurotransmitter:** Choline is taken up into the neuron and combined with coenzyme A (CoA) in the presence of **choline acetyltransferase** to form acetylcholine.

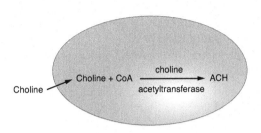

FIGURE 5.4

    i. It has to be synthesized in the presynaptic structure and is released in response to action potentials.
    ii. The action potential causes Ca^{++} to come in and vesicles to go to the neurolemma and exocytose ACH to the synapse (review muscle physiology).
    iii. Neurotransmitter has to bind to receptors.
    b. Mechanism of release: **1, 2, 4, and 5 are cholinergic; postganglionic sympathetic is not.**

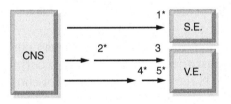

FIGURE 5.5

    i. Description & density of receptors
    ii. **Responses in postsynaptic structure**
        (1) #1 is somatic efferent
        (2) #2 is preganglionic sympathetic efferent
        (3) #3 is postganglionic sympathetic efferent
        (4) #4 is preganglionic parasympathetic efferent
        (5) #5 is postganglionic parasympathetic efferent
    iii. **Receptor types: ACH** acts like a "master key." **Nicotine** binds and affects certain receptor types. **Muscarine** binds and affects certain receptor types.

(1) Nicotinic: These receptors respond to nicotine but not to muscarine; N1 and N2 are two subtypes of receptors:
   a. N1: Always excitatory; works by altering the ionic permeability of the postsynaptic site and increasing the Na^+ permeability
   b. N2: Always excitatory; works like N1
(2) Muscarinic: These respond to muscarine but not nicotine. Five subtypes of receptors:
   a. M1: Typically excitatory; generally works via phospholipase C system to communicate across the cell membrane
   b. M2: Often inhibitory; inhibits cAMP and/or increases the K^+ permeability
   c. M3: Like M1
   d. M4: Like M2
   e. M5: Like M1 and M3

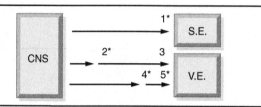

FIGURE 5.6

(3) #1 is somatic efferent: N2 receptors
(4) #2 and #4: N1 receptors: always excitatory
(5) #5: M1, M2, and M3 receptors are on smooth or cardiac muscle or a gland: visceral effectors
   a. In some places excitatory and in some places inhibitory
   b. Heart rate under parasympathetic influence decreases due to stimulating M2 receptors
   c. Parasympathetic influence on GI system increases motility due to M1 and M3 receptors.
(6) There are more muscarinic receptors than nicotinic receptors.
(7) #3 does not have ACH receptors because it is not cholinergic.
  c. Destruction of ACH
    i. Effect of ACH on the receptor is short lived (only a few milliseconds) because of an enzyme called **acetylcholinesterase.**
    ii. Acetylcholinesterase can destroy only ACH that is **free in the synaptic cleft.**
    iii. ACH that is bound to receptors is in equilibrium with the ACH that is in the synaptic cleft.

V. Differences in neurotransmitters and receptors: Adrenergic (Epinephrine and Norepinephrine Secretors); #3 in Figure 5.6—Sympathetic postganglionic
  a. Anatomic location
    i. Postganglionic sympathetic neurons release 95% norepinephrine and 5% epinephrine.
     (1) Not all adrenergic; some sympathetic fibers innervate certain **sweat glands** and are cholinergic.
      a. These are still sympathetic because of the short preganglionic and long postganglionic fibers, and distribution is via T1 to L3 spinal nerves roots.
    ii. Adrenal medulla: This is a gland (although it behaves more like nervous tissue) and is unusually innervated.

(1) **Composed of two parts:** inner part is the medulla and outer part is the adrenal cortex (which secretes adrenal cortical hormones).
(2) There is just a preganglionic fiber. The medulla can be thought of as a postganglionic fiber in a ball or bundle.
(3) It secretes/produces **~20% norepinephrine and ~80% epinephrine.**
(4) The adrenal medulla dumps norepinephrine and epinephrine into the blood; there is no post-sympathetic neuron.
(5) So not really a neurotransmitter—called a **neurohormone.**
(6) What kind of receptor are on the adrenal medulla? **N-1**
b. Synthesis of norepinephrine:

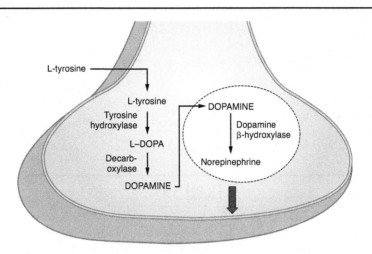

FIGURE 5.7

i. Transport L-tyrosine into the neuron; in the presence of an enzyme (tyrosine hydroxylase), L-DOPA is made. Then in the presence of the enzyme decarboxylase, dopamine is made.
c. **Storage:** Dopamine is taken into vesicles, and then in combination with **dopamine beta-hydroxylase,** we synthesize norepinephrine.
d. **Release:** Ca^{++} comes in and the vesicle goes to the membrane to exocytose norepinephrine to the synapse.
e. Binding to receptors
    i. **Types**
        (1) **Alpha receptors** love epinephrine more than norepinephrine and more than isoproterenol (ISO) **(E > NE > ISO).**
        (2) **Beta receptors (ISO > E = NE)**
    ii. **Generalizations**
        (1) Alpha receptors are generally associated with excitation: **Alpha-1 (phospholipase C mechanism)**
            a. **Alpha-2** are generally inhibitory (cAMP).
        (2) Beta receptors are generally inhibitory: **Beta-2 (cAMP).**
            a. **Beta-1, Beta-3 are generally excitatory** (cAMP).

So, if excitation, we would *expect* α-1; if inhibition, we would *expect* β-2. As we will see, there will be exceptions to these general rules.

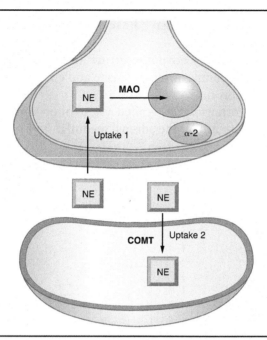

FIGURE 5.8

i.   **Uptake 1 (by adrenergic terminal: deactivation by uptake 1)**
   (1) Norepinephrine is taken up from the synapse back into the postganglionic sympathetic neuron.
   (2) Within the cytoplasm of the neuron is an enzyme (a monoamine oxidase; MAO) that can break down norepinephrine.
   (3) MAO is present in the cytoplasm (on mitochondria) but not in the vesicles.
   (4) Norepinephrine can be transported into a vesicle to protect it from MAO
   (5) Example: Could have a drug that blocks the uptake of norepinephrine into vesicles (Reserpine/SERPASIL). This would decrease sympathetic activity.
ii. **Uptake 2:** Some norepinephrine gets into the post-sympathetic structure by uptake 2.
   (1) Catechol-O-methyltransferase **(COMT):** An enzyme in the postsympathetic structure that destroys norepinephrine.
iii. **Alpha-2 receptors:** On postganglionic sympathetic neurons are alpha-2 receptors, which are sensitive to norepinephrine.
   (1) Norepinephrine binds to alpha-2, and the exocytose mechanism is blocked.
   (2) Alpha-2 binds to norepinephrine and stops the vesicles from being exocytosed.
   (3) The alpha-2 receptors are presynaptic!

Side note: Nonclassic Neurotransmitters of the Autonomic Nervous System
   Some neurons exhibit **co-localization** (secrete more than one neurotransmitter).
   ATP is released from some postganglionic sympathetic neurons to cause vasoconstriction.
   NO is released in some places to cause vasodilation.
   These are co-localized with ACH and/or norepinephrine.

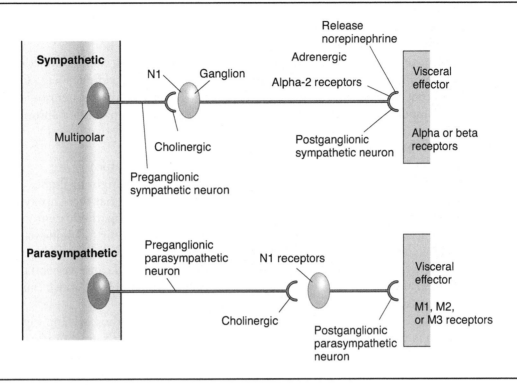

FIGURE 5.9

VI. Differences in Function
  a. General Afferent and Efferent

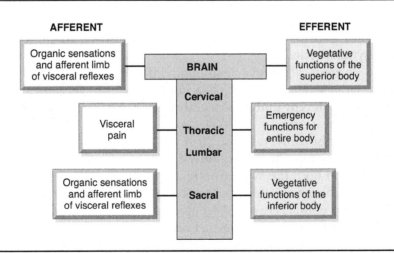

FIGURE 5.10

With regard to the figure above, note that when pertaining to humans *per se*, the "anterior parts of body" would most accurately be called "superior parts of the body" or "cranial end," whereas "posterior parts of body" would most accurately be called "inferior parts of the body" or "caudal end."

i. **Cranial nerves**
   (1) **Efferent** (parasympathetic because a cranial nerve): Promotes vegetative functions, such as resting or digesting, pertaining to the superior parts of the body.
   (2) **Afferent:** Organic sensations in the superior part of the body, such as thirst, hunger, or nausea. Provides the *highway* for afferent **limb of visceral reflexes**: e.g., GI activity (always runs *with* parasympathetic pathway). Note that this afferent neuron is a typical pseudounipolar afferent neuron that travels *with* the parasympathetic neuron.
ii. **Cervical**
   (1) Cervical spinal nerves **have only somatic function.**
iii. **Thoracic and Lumbar**: sympathetic function (somatic function, too)
   (1) **Efferent:** Promotes emergency functions to the entire body; fight or flight.
   (2) **Afferent:** Involved in the perception of **visceral pain. Note** that these are typical pseudounipolar afferent neurons that now travel *with* the sympathetic neurons.
      a. Examples: stomach ache, pain of menstrual cramps, childbirth, angina (chest pain)
      b. Tissue destruction does not cause visceral pain (like it does for somatic pain); smooth muscle contractions or reduced blood flow (ischemia) will cause visceral pain.
      c. **Visceral pain is often referred to the soma**. Example: Heart attack causes pain in the chest wall, shoulder, arm (somatic structure).
         Examples: childbirth, labor pains (the pain is referred to the back and vagina)
      d. **Pain is typically referred to a specific dermatome:** first four thoracic spinal nerves each receive fibers from the heart; pain is referred because afferent impulses are sent to the pain center in the brain. The visceral fibers also go to the pain center, and **when impulses arise there, most people are more apt to feel somatic pain than visceral pain.**
      e. The brain thinks that T1–T4 impulses must mean somatic pain and not visceral pain, because the brain usually receives more somatic impulses than visceral impulses. **Think: You have spent *years* interpreting pain there as somatic pain; not surprising that when visceral pain arises there is spillover to somatic pain areas *or* it is interpreted as somatic pain.**
      f. Evidence: People who have lots of visceral pain get better at localizing it.
iv. **Sacral: Parasympathetic**
   (1) **Efferent:** Promotes vegetative functions in inferior parts of the body.
   (2) **Afferent:** Organic sensations such as a **full bladder, full rectum,** and **sexual sensations**; also the **afferent limb of visceral reflexes,** such as **sexual reflexes, urination, and defecation**

b. **Tonic activity and double innervations: Most visceral effectors are doubly innervated (sympathetic and parasympathetic). In most cases there is tonic activity (low-frequency continuous activity) in both the sympathetic and parasympathetic pathways.**
   i. Because of the double innervations and tonic activity, we have lots of choices for changing or regulating the rates of things (we can "let up on the brake" *and* "press the gas").
   ii. Example: **Duodenal motility** is associated with "rest and digest." Parasympathetic activity increases duodenal activity, whereas sympathetic activity decreases it.
   iii. Example: **Increase in HR**: sympathetic would increase (β-1) and parasympathetic would decrease (M2).
   iv. **Sympathetic speeds up activity in an emergency but inhibits activity during "rest and digest."**
   v. **Parasympathetic increases "rest and digest" activity but inhibits "fight or flight" activity.**

c. **Predictions:** Example: Secretion of procarboxypeptidase (a **pancreatic digestion enzyme**). Parasympathetic increases activity and sympathetic decreases activity.
What kind of cholinergic receptors are on these acinar cells that produce procarboxypeptidase?

d. **Particular organs**
   i. **Eye: Size of the pupil:** controlled by the iris (surrounds the pupil). Two muscle networks innervate the iris.
      (1) Circular muscle: When it contracts, it makes the pupil get smaller/constrict.
      (2) Radially located muscle: Contracts and pulls the iris back or dilates the pupil.
      (3) Pathway to the circular muscle is parasympathetic (M1), and that to the radial muscle is sympathetic (α-1).
      (4) Stimulus causing the pupil to constrict is light; if there is no light, then stop parasympathetic stimulation (specific).
      (5) Dilating the pupil is due to sympathetic stimulation (nonspecific). Also, if you inhibit parasympathetic stimulation, then the pupil will dilate.
   ii. **Eye: Accommodation of lens of eye:** allows the lens to change shape to focus on near or far objects appropriately.
      (1) Ciliary muscle: Parasympathetic stimulation makes the muscle contract and causes the lens to bulge outward, allowing focus on near objects **(M1, M3). Sympathetic stimulation** allows the muscle to relax; lens flattens some, can focus on far objects (β-2).
      (2) Presbyopia ("elder eyes"): Lens begins to lose elasticity, so instead of being flexible, the lens hardens and becomes inflexible. There is **nothing wrong with the ciliary muscle,** but rather the lens will not bend—thus the patient has trouble seeing near objects.
         a. When the ciliary muscle contracts and changes the shape of the lens, it also enhances the drainage of aqueous humor; too much aqueous humor causes a buildup of pressure, known as **glaucoma**; if you stimulate the ciliary muscle, then you allow for the drainage of aqueous humor.
         b. There are M1 and M3 receptors here, so you may continuously stimulate M1 and M3 receptors to drain aqueous humor. You may also block the β-2 receptors.
   iii. **Heart:** Receives sympathetic and parasympathetic innervation.
      (1) Sympathetic: From the first four thoracic spinal nerves (T1–T4)
         a. Sympathetic efferent sends transmissions to the heart. Norepinephrine is what is predominantly released onto the heart.
         b. Sympathetic innervation increases HR, increases the force of contraction (stroke volume) and decreases conduction time (β-1 receptors are found on the myocardium). Overall, this increases cardiac output.
      (2) Parasympathetic: Decreases HR. Cholinergic receptors are M2. This decreases HR by the left vagus, innervating the AV node and atrial musculature, and the right vagus, innervating the SA node and the atrial musculature. There is not much parasympathetic effect on force (mostly effects are on HR).
   iv. **Arterioles:** Located between the capillaries and the major blood vessels; regulate blood flow
      (1) Usually there are **either alpha or beta receptors.**
      (2) Arterioles are unusual and typically have only sympathetic innervation.
      (3) Alpha-1 receptors cause the arteriole smooth muscle to contract, which causes vasoconstriction.
      (4) Beta-2 receptors cause the muscle to relax, which causes vasodilation.
      (5) There are more alpha receptors than beta receptors, so the overall response is vasoconstriction.

(6) Exceptions: Arterioles supplying the **skeletal muscle and also the coronary arteries of the heart**; there are more β-2s than α-1s.

(7) Under sympathetic duress (times of emergency) you get vasoconstriction except to the heart, skeletal muscle, and liver, because you need blood flow here. **Note** that if you really dilated all muscle vascular beds at once, you would pass out; when exercise causes an increase in sympathetic outflow, the initial response is widespread constriction, followed by dilation of *working* muscle tissue vasculature, thought to be due to local factors (e.g., H^+, adenosine, etc.). In this manner BP is maintained and working muscles receive the needed blood flow.

(8) Arterioles do have muscarinic receptors that cause vasodilation (M2), but these receptors **are not innervated**; however, the receptors can be turned on pharmacologically; exceptions to this are the arterioles supplying the erectile tissues, **which are doubly innervated by sympathetic and parasympathetic fibers** (penis and clitoris). The parasympathetic innervation is via M3s, which cause release of NO to dilate the arterioles locally.

(9) Also, some arterioles that supply skeletal muscle have long preganglionic sympathetic fibers that directly innervate their muscarinic receptors (M2). So mostly β-2s are responsible for the increased blood flow to muscle under duress (remember, the circulation epinephrine in an emergency would not stimulate the M2s).

What happens if we inject an animal with ACH?

Would this ACH have any added effect on tonic adrenergic stimulation at places such as the GI tract?

Hint: Think about presynaptic structures of the postganglionic adrenergic neurons.

v. **Lungs: Diameter of the bronchi** and **secretion of bronchiole glands**

    (1) Sympathetic: Causes bronchi to dilate and glands to cease in their secretions. Allows **open and dry bronchi**; an inhibitory response—**β-2 receptors.**

    (2) Parasympathetic: Causes constriction of bronchi and secretion of glands. **M1 and M3 receptors. War gases** may stimulate these. **Asthma medication** may stimulate β-2s or inhibit M1, M3.

vi. **Lacrimal and nasopharyngeal glands:** Lacrimal = tear-producing gland above eye; nasopharyngeal = mucus-producing gland; typically they have only **parasympathetic innervation and M1 receptors**.

    (1) Decongestants: **Neosynephrine (or pseudoephedrine**—the thing that makes Claritin-D "work") is a potent alpha-1 stimulator. Alpha-1 receptors cause vasoconstriction, and the blood flow to the nasopharyngeal glands slows and mucus secretion stops. **No sympathetic innervation** to these glands!

vii. **Urinary bladder:** Innervated by sympathetic and parasympathetic fibers
    (1) For urination to occur, the **detrussor muscle has to contract (M1)** and the **sphincter of the bladder has to relax (M2).**
    (2) **Sympathetic:** Detrussor muscle relaxes (β-2) and sphincter contracts (α-1) and **inhibits urination.** *[handwritten: β-3]*
    (3) **Parasympathetic:** Favors urination. The detrussor muscle contracts and the sphincter relaxes.
    (4) **Drugs: DUVOID**: A strong muscarinic stimulator; for trouble urinating (after surgical anesthesia, some have **nonobstructive urinary retention**).
    (5) **Anxiety in public restrooms** may cause sympathetic output and inhibit urination.
viii. **Pancreas**
    (1) **Exocrine pancreas:** produces digestive enzymes.
      a. **Parasympathetic:** stimulates secretion
      b. **Sympathetic:** inhibits secretion
    (2) **Endocrine pancreas:** hormone producer
      a. **Parasympathetic:** excitatory; increases insulin secretion
      b. **Sympathetic:** inhibits insulin secretion
ix. **Gastrointestinal tract:** motility, sphincters, and secretions (think, in general, β-2 and M1, M3).
    (1) **Parasympathetic:** increases motility, causes sphincters to relax, and increases secretions
    (2) **Sympathetic:** inhibits motility, causes the sphincters to contract, and inhibits secretions
x. **Salivary glands**
    (1) **Parasympathetic:** excitatory; increases volume of salivation (a copious, watery secretion)
    (2) **Sympathetic:** causes a thick secretion but is very minimal (a much reduced rate). This secretion is high in digestive enzymes. Although this intuitively seems strange—snakes have poison glands that are simply advanced salivary glands—these provide protection in an "emergency."
xi. **Piloerector muscles:** cause the hair to stand up, particularly on the appendages; smooth muscle at the base of the hair—when stimulated the hair stands up.
    (1) **Only sympathetic innervation**
    (2) **Piloerection:** hair follicles stand up
    (3) Hair used for thermal regulation
    (4) Chill bumps or **"horripilate"**; fear causes chill bumps as well
xii. **Eccrine sweat glands:** innervated **only by the sympathetic nervous system**. Unusual because they have cholinergic receptors on their surface (M1 receptors). These are still sympathetic, though, because they arise from T1–L3 and have a short preganglionic fiber and long postganglionic fiber.
    (1) **The soles of the feet and hands are adrenergic receptors like they are supposed to be.**

    When you begin an exam and get flustered and nervous (emergency response), what parts of your body sweat?

    Arm? Hands? Back? Axillary regions? Pubic area? Feet?

xiii. **Apocrine sweat glands:** Under the arms and in the pubic region (also in pigmented part of breast).

    (1) These begin activity during puberty and secrete a thick, odiferous secretion.

    (2) Adrenergic fibers go to them, and they have α-1 receptors.

    (3) Sympathetic: sweat on the palms of hands and feet from eccrine sweat glands and underarms from apocrine sweat glands

xiv. **Penis**

    (1) Sympathetic: ejaculation

    (2) Parasympathetic: erection

xv. **Kidney:** β-1 receptors to enhance production of renin; renin is secreted to increase BP (via vasoconstriction).

xvi. **Fat**

    (1) Sympathetic stimulation of β receptors causes an increase in lipolysis.

Draw each visceral organ mentioned (with receptors and innervations).

You will need to be able to do this to play the drug game coming up.

Eye

Heart

Arterioles

Lungs

Lacrimal and nasopharyngeal glands

Urinary bladder

Pancreas

GI tract

Salivary glands

Piloerector muscles

Eccrine sweat glands

Apocrine sweat glands

Penis

Kidney

Fat

VII. Differences in Pharmacology: Pharmacologically, we will somehow perturb a physiological mechanism. For this course we will memorize several drugs and types of drugs so that we can better understand how these physiological mechanisms work and how they can be modulated accordingly.

  a. **Cholinergic:** Overall there are more muscarinic receptors, so there is a visceral effect on Ms if we stimulate cholinergic receptors.

   i. **Cholinomimetics:** drugs that mimic ACH; it is hard to give ACH to patients who need it because of ACH esterases. ACH esterases are very active in breaking down ACH. Cholinomimetics fool the receptor, but not ACH esterases (i.e., the receptor binds them and acts exactly as if it were ACH, but the ACH esterases do not break it down like they would ACH). This effectively increases the half-life of ACH.

   (1) Clinical uses
      a. **Glaucoma:** Predominantly, cholinomimetics are used to treat this—they stimulate the ciliary muscle to drain aqueous humor.
      b. **Urinary retention**
      c. Tobacco substitute
      d. **Xerostomia:** Perpetual dry mouth due to a lack of saliva; muscarinics cause salivation.

   (2) Examples
      a. **Pilocarpine (ISOPTO-CARPINE):** N and M stimulator; used to treat glaucoma
      b. **(DUVOID):** M stimulator; used to treat urinary retention
      c. **Nicotine (NICORETTE, NICODERM):** N stimulator generally used to help stop smoking; stimulates nicotinic sites. Has biphasic nature—at low doses it stimulates N-1s, but at high doses it inhibits N-1s.

   ii. **Receptor blockade: Nicotinic-2**
   (1) Clinical uses
      a. Intubations
      b. Muscle relaxation
   (2) Examples
      a. **Curare (TUBADIL):** N2 block
      b. **Succinylcholine (ANECTINE):** N2 block; allows one depolarization and then blocks
      c. Inhibit contraction of skeletal muscle. Must be careful because **respiratory musculature is also skeletal muscle.**

(3) **Receptor Blockade—Nicotinic-1**
    a. **Clinical uses**
        i. Hypertensive crisis
    b. **Example**
        i. **Trimethaphan (ARFONAD)** (N-l)
        ii. **Nicotine:** Once used as antihypertensive because it blocks N1 receptors in adrenal medulla and ganglia, but there are many side effects.
(4) **Receptor blockade: Muscarinic:** Blocks postganglionic parasympathetic sites—less of a parasympathetic effect.
    a. **Clinical uses**
        i. **Preop** medication: to open and dry airways
        ii. **Eye exam:** Eye drops with a muscarinic blocker are used to allow light on the eye, with no pupil constriction.
        iii. **Bladder instability**: block muscarinic receptors (because they make the detrussor muscle contract and the sphincter relax) to maintain a more stable bladder.
        iv. **GI:** used to block M stimulation of gut to decrease motility
        v. **Motion sickness:** central mechanism
        vi. **Bronchiospasm:** open and dry bronchi when M1 and M3 of lungs are blocked
    b. **Examples**
        i. **Atropine (Bella Donna,** "beautiful woman"): naturally occurring; causes the eyes to dilate. Not used today because it takes hours to wear off.
        ii. **(BENTYL):** used in patients with irritable bowel syndrome (IBS); decreases GI motility.
        iii. **(DETROL):** used to treat bladder instability
        iv. **(MYDRIACYL):** used in eye exams
        v. **(ATROVENT):** treats asthma; not receptor blocker. Anticholinergic—ACH can still bind, but interferes with the message
(5) **Anticholinesterases:** prolong the life of ACH (same uses as cholinomimetics)
    a. **Clinical uses**
        i. **Myasthenia gravis:** a somatic problem; neuromuscular junction disease in which immune system inappropriately attacks ACH receptors (N-2s). So we use this to keep ACH out in the cleft longer—keep ACH there longer to continually stimulate the receptors that are there.
        ii. **Postoperative urinary retention**
        iii. **Glaucoma**
        iv. **Insecticide**
        v. **War gases:** Cause bronchi to contract and secrete fluid, so you will eventually drown in your own secretions.
    b. **Examples**
        i. **Physostigmine (ESERINE)**
        ii. **(MESTINON)**
        iii. **Organophosphates (PARATHION, DIAZINON, MALATHION):** persistent anticholinesterases
        iv. **SARIN:** active ingredient in war gases
b. **Adrenergic Sites** (postganglionic sympathetic sites)
    i. **Block transmitter synthesis (DEMSER):** block the synthesis of neurotransmitter (epinephrine or norepinephrine) by blocking the enzyme **tyrosine hydroxylase (catalyzes L-tyrosine → L-Dopa)**
    (1) Decrease sympathetic effect—once used to treat hypertension

ii. **False transmitter** (methyldopa or **ALDOMET**)
    (1) Methyldopa (instead of L-Dopa) makes alpha methyl norepinephrine, which is an adrenergic drug and has no effect on alpha-or beta-1 receptors, but ***does* affect alpha-2 receptors.**
    (2) Presynaptic receptors (α-2) are inhibitory and shut off adrenergic receptors and sympathetic function.
    (3) Antisympathetic; used to treat hypertension

iii. **Block release (ISMELIN):** block the release of norepinephrine—subsequently, in the presence of an action potential, nothing much happens
    (1) Used to treat hypertension

iv. **False displacement** (amphetamines, e.g., ephedrine [**DEXEDRINE**]): **Displacement of the neurotransmitter without an action potential** = more sympathetic function. These **also directly stimulate α and β receptors.**
    (1) Used during surgical procedures for hypotension.
    (2) Amphetamines are used in weight-control drugs (addictive) and cause you to get "really pumped up."
    (3) Used to treat narcolepsy.
    (4) Some effects are centrally mediated.

v. **Block transport of norepinephrine back into storage granules (reserpine (SERPASIL)):** Reserpine inhibits the transport back into the vesicles that hold norepinephrine. Thus, MAO can destroy the norepinephrine.
    (1) Decreases sympathetic effect.
    (2) Once used for hypertension, but **unpleasant side effects (depression)**

————**The adrenergics above involve presynaptic mechanisms.**—————-

vi. **Adrenomimetics**: Alpha agonists
    (1) Clinical uses—Alpha-1: Promote sympathetic effects in the organ systems
        a. **Shock:** Too little blood or too much vasodilation, such that pressure falls really low; treat arterioles and increase vasoconstriction in veins and spleen; shock can occur in hypotension from spinal anesthesia.
        b. **Hypotension**
        c. **Nasal decongestants**
        d. **Local anesthetic adjunct:** Partner in local anesthesia, because it can control spread of drug due to vasoconstriction (also controls bleeding).
    (2) Clinical uses—Alpha-2: hypertension: inhibitory
    (3) Examples
        a. **Epinephrine, adrenaline:** Alpha-1 & -2; Beta-1 & -2
        b. **Norepinephrine (LEVOPHED):** Alpha-1 & -2 (Beta-1 & -2)
        c. **Phenylephrine (NEOSYNEPHRINE):** Alpha 1: a nasal decongestant
        d. **(EPIPEN):** Alpha-1 (cardiac shock, allergic reactions)
        e. **Clonidine (CATAPRES)** Alpha-2: treats hypertension
        f. **(ALPHAGAN):** Alpha-2 agonist; decreases aqueous humor

vii. **Adrenomimetics**: Beta agonists
   (1) Clinical uses
   a. **Beta-1**: heart—increase cardiac output
      Treat
      —cardiac arrest
      —heart failure
      —shock
   b. **Beta-2**
      —Asthma: treatments cause the bronchi to dilate and the glands to stop secreting.
      —Delay labor: effect on the uterus (decrease myometrial activity)
   (2) Examples
   a. **Isoproterenol (ISUPREL)**: Beta-1 & -2; treat shock and affects the bronchi.
   b. **(DOBUTREX)**: Beta-1
   c. **(PROVENTIL) (MAXAIR)**: Beta-2; used to treat asthma
   d. **(YUTOPAR)**: Beta-2; delays premature labor of the uterus
viii. **Antagonists**
   (1) Alpha blockers
   a. **Clinical uses**
      —**Hypertension**: blocks alpha receptors of the arterioles, causing vasodilation and decreased BP
      —**Pheochromocytoma**: A disease that encompasses benign tumors on the adrenal medulla. This produces norepinephrine and epinephrine in excessive amounts. Pronounced sympathetic effect and high BP are the results.
      —**Benign prostatic hyperplasia**: Prostate gland (located by the urethra at the bladder) grows bigger (hyperplasia) as men age. With an enlarged prostate, it is difficult to urinate.
   b. **Examples**
      —**Phentolamine (REGITINE)**: Alpha-1 & -2 blockers (pheochromocytoma)
      —**(HYTRIN)**: Alpha-1; antihypertensive
      —**(MINIPRESS)**: Alpha-1; antihypertensive
      —**(CARDURA)**: Antihypertensive
      The above three have a side effect of orthostatic hypotension; regulation of BP during positional change (standing up) is problematic. Gravity pulls blood down to feet; alpha-1 blockers disable quick vasoconstriction upon standing, affecting maintenance of adequate brain perfusion.
      —**(FLOMAX)**: Used to treat prostatic hyperplasia
   (2) Beta blockers
   a. **Clinical uses**
      —**Hypertension**: Predominantly affect heart; decrease HR and force of contraction and thus decrease BP
      —**Angina**: The pain associated with not enough perfusion at the heart muscle; beta blockers can decrease the demand on the heart and thus relieve angina.
      —**Dysrhythmias**
      —**Migraine**: No clear mechanism established
      —**Glaucoma**: Ciliary muscle doesn't have a lot of sympathetic innervation; but if you block sympathetic innervation (β-2), then aqueous humor drainage improves.

    b. **Examples**
       —**Propranolol (INDERAL):** Beta-1 & -2; used for hypertension; **generalized beta blocker (nonspecific).**
       Would you give an asthmatic patient with hypertension INDERAL?

       —**(TIMOPTIC):** Beta-1 & -2; used to treat glaucoma.
       <u>**Good antihypertensive drugs:**</u>
       —**(LOPRESSOR):** Beta-1
       —**(TENORMIN):** Beta-1
       —**(TOPROL-XL):** Beta-1

  (3) **Alpha & Beta combinations**
    a. **(COREG):** Beta-1 and -2, plus alpha-1 blocker. Alpha-1 is working at the vasculature to cause dilation, and beta blockers are working at the heart.

SUGGESTED READINGS

1. Guyton, A.C., and Hall, J.E. 2011. Chapter 60 in *Textbook of medical physiology*, 12th ed. Baltimore: Saunders.
2. Koeppen, B.M. and Stanton, B.A. 2008. Chapter 11 in *Berne & Levy Physiology*, 6th ed. St. Louis: Mosby.

THE DRUG GAME

Rules: Assume an isolated preparation (this means we exclude the preganglionic sympathetic site but include the preganglionic parasympathetic site) and exclude postganglionic presynaptic sympathetic receptor site (α-2). Only postsynaptic receptors will be included in game. Also, assume tonic activity in organs with both sympathetic and parasympathetic pathways. And most importantly, follow the drug sequence rule: that is, the effect of the second drug is compared to the state of the organ *before* addition of the *second* drug, *not* overall. There will be many of these questions on Exam II, so know how to play with these! Memorizing the drugs and the visceral organs is just the beginning.

1. What is the effect of pilocarpine on the rate of contraction of the heart?

2. What is the effect of atropine on HR?

3. What is the effect of atropine plus pilocarpine on HR?

4. What is the effect of physostigmine plus ACH on the heart?

5. What is the effect of ACH plus atropine on HR?

6. What is the effect of Inderal plus epinephrine on the diameter of the bronchi?

7. What is the effect of Tenormin plus epinephrine on the diameter of the bronchi?

8. What is the effect of Tenormin plus ACH on the diameter of the bronchi?

ANSWERS TO THE ABOVE DRUG GAME

1. Decrease
2. Increase
3. No effect
4. Decrease
5. Increase
6. No effect
7. Increase
8. Decrease

If these answers seem odd, be sure to go back and review the rules of the game.

CHAPTER 5 QUESTIONS

1. _____ The parasympathetic nervous system
   a. Is distributed in cranial nerves
   b. Is distributed in cervical spinal nerves
   c. Is distributed in thoracic spinal nerves
   d. All of these
   e. Two of these

2. _____ Not associated with the sympathetic nervous system
   a. Superior mesenteric ganglion
   b. N-1 receptors
   c. Pyramidal pathway
   d. All of these
   e. Two of these

3. _____ The fourth thoracic spinal nerve
   a. Has short preganglionic fibers in it
   b. Has pseudounipolar neurons in it
   c. Has the dorsal column system "running" in it
   d. All of these
   e. Two of these

4. _____ Muscarinic receptors
   a. Are always excitatory
   b. Are more numerous than nicotinic receptors
   c. Are found associated with the sympathetic nervous system
   d. All of these
   e. Two of these

5. _____ In a cervical spinal nerve, there could be impulses
   a. Causing the heart to slow
   b. Sensing visceral pain
   c. Sensing thirst
   d. None of these
   e. Two of the above

6. _____ Used to treat hypertension
   a. An alpha-2 agonist
   b. A muscarinic blocker
   c. A beta-1 agonist
   d. All of these
   e. Two of these

7. _____ Could be effectively used to treat glaucoma
   a. A cholinomimetic
   b. A beta blocker
   c. ALPHAGAN
   d. All of these
   e. Two of these

8. _____ The pyramidal pathway
   a. Has a short preganglionic fiber in it
   b. Has a long preganglionic fiber in it
   c. Has a multipolar neuron in it
   d. None of these
   e. Two of these

For questions 9–19, choose one of the following:
   A. Increases or is greater than
   B. Decreases or is less than
   C. Has no effect or is equal to

9. _____ The number of paravertebral ganglia, as compared with the number of spinal nerves with a sympathetic function

10. _____ The length of the postganglionic sympathetic fiber, as compared with the length of the postganglionic parasympathetic fiber

In the following questions, assume isolated preparations.
11. _____ The effect of Timoptic on HR

12. _____ The effect of Cardura on HR

13. _____ The effect of Dobutrex on intestinal motility

14. _____ The effect of Isopto-Carpine on the detrussor muscle

15. _____ The effect Tubadil plus epinephrine on piloerection

16. _____ The effect of atropine on the production of sweat by the eccrine sweat glands of the hands

17. _____ The effect of Inderal plus Neosynephrine on the diameter of the bronchi

18. _____ The effect of Yutopar plus Tenormin on HR

19. _____ The effect of Hytrin plus pilocarpine on the diameter of an arteriole

CHAPTER 6

Cardiovascular Physiology I

I. **Introduction**
   a. General Functions
      i. Important **transport system**
         (1) If you want to change the composition of any of the fluid compartments, you start at the plasma.
      ii. Important role in the **immune system**—white blood cells
      iii. Important role in **acid-base balance**
   b. Components
      i. **Pump (heart)**
      ii. **Vessels** (arterioles, capillaries)
      iii. **Blood**
   c. Divisions: Two complete circuits
      i. **Pulmonary:** heart to lungs and back to heart
         (1) Much shorter circuit
         (2) The right side of the heart
      ii. **Systemic:** heart to body and back to heart
         (1) The left side of the heart

II. **Heart: Two major components—The contractile portion** generates force (cardiac muscle), while the **conductile portion** initiates electrical events and disperses them across the heart muscle (SA node, AV node, the Bundle of His, Purkinje System, and the Bundle Branches).
   a. **Microanatomy:** The cardiac muscle is highly branched and striated with dark bands that mark the junctions (boundaries) of the adjacent cardiac muscle fibers, called intercalated disks.
      i. **Intercalated disks:** Boundaries are **areas of low electrical resistance.** An action potential propagated in the cardiac muscle fiber is propagated to the adjacent cardiac muscle fibers (via gap junctions).
      ii. This is very different from skeletal muscle. There is no reason for the cardiac muscle fibers to be privately innervated, since it has these areas of low electrical resistance.
      iii. Troponin, tropomyosin, actin…these are generally the same in skeletal and cardiac.

iv. Sub-sarcolemma Ca^{++} stores: Cardiac muscle has a **sarcoplasmic reticulum** (SR) to hold Ca^{++}. This SR is better than smooth muscle but worse than skeletal.

v. **Cardiac muscle has T-tubules that are very well developed** (~5X diameter of skeletal muscle). This T-tubule system has abundant mucopolysaccharide, called glycocalyx, that stores Ca^{++} and can deliver it into the cell at the appropriate time.

vi. Cardiac cells also have a **high density of mitochondria** (cardiac muscle fibers have the highest metabolic demand per gram of any tissue in the body).

SR Complexity: Skeletal > Cardiac > Smooth

T-Tubules: Cardiac > Skeletal > Smooth (no T-tubules)

b. Electrical Events

    i. Fast Response

        (1) Description: Atrial muscle and ventricular muscle are contractile fibers (Purkinje fibers are *not* contractile fibers but rather are derived from ventricular tissue and have a similarly shaped action potential).

           a. Have a resting membrane potential (-90 to -95 mV) that is more polarized than skeletal muscle

           b. 400 milliseconds: longer than action potential in the skeletal muscle.

        (2) Ionic events

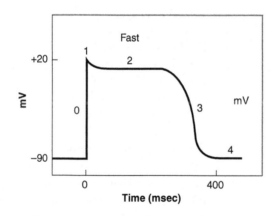

FIGURE 6.1

    a. **At 0:** What is responsible for the rapid depolarization? Rapid increase in Na^+ permeability.

    b. **At 3:** Repolarization due to decline in Na^+ permeability and an increase in K^+ permeability.

    c. **Depolarization/Repolarization so far is no different than usual action potential.**

    d. **What causes the plateau?** Due to the entry of Ca^{++} from the slow Ca^{++} channels (mainly from the aforementioned glycocalyx).

    e. This action potential is found in atrial and ventricular muscle, and Purkinje fibers. Cardiac muscle fibers are absolute refractory for a long time, and thus you cannot tetanize cardiac muscle fibers (which is good!). This increased Ca^{++} is also likely to enhance the contractile mechanism. We can increase stroke volume (SV) by increasing Ca^{++} influx. **Calcium channel blockers,** as we will see, will work to decrease SV and thus decrease BP. Epinephrine and norepinephrine on the heart enhance the Ca^{++} delivery through the slow Ca^{++} channels of the heart (and also act on increasing HR).

  ii. Slow Response

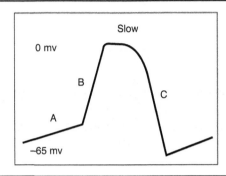

FIGURE 6.2

(1) Description: response in the **SA node** (the pacemaker of the heart) and the **AV node** (between atrial and ventricle) of depolarization and repolarization
   a. At rest it is not nearly as polarized as the fast contractile fibers. It barely goes to 0 millivolts before it repolarizes.
   b. **No firm resting membrane potential! In fact,** part A in Figure 6.2 is really called a prepotential, not a resting membrane potential.
   c. In Figure 6.2, **B is depolarization and C is repolarization.**
(2) **Ionic events**
   a. **A:** Membrane's permeability to K^+ progressively declines (if the membrane was impermeable to K^+ then it would be at 0 millivolts). Progressively less permeability to K^+ causes the voltage to approach 0. Some Na^+ also leaks in. Note that in the literature this is referred to as the "funny current" channel (I_f).
   b. **B:** Predominantly due to progressive movement inward of Ca^{++} through slow Ca^{++} channels
   c. **C:** Predominantly due to the outward movement of Ca^{++} through slow Ca^{++} channels
c. **Excitation Coupling:** This is only slightly different than in skeletal muscle. Still "all or none" contraction of a fiber.
   i. Depolarization of the sarcolemma and T-tubules: The action potential depolarizes the sarcolemma and T-tubules and causes the further release of Ca^{++} from the T-tubules (plateau: Ca^{++} coming into the cell from the T-tubules—these DHP receptors are a type of Ca^{++} channel sensitive to voltage changes). (Remember that Ca^{++} is coming in from extracellular fluid as well as the SR.)
   ii. Extracellular Ca^{++} (or **trigger Ca^{++}**) induces more Ca^{++} from the SR to be released.
   iii. **In the skeletal muscle,** the SR releases just enough Ca^{++} for the troponin binding sites and has the maximum effect possible (i.e., more Ca^{++} would *not* increase the strength of

contraction). **But in cardiac muscle, additional calcium from the glycocalyx can and does increase the strength of a contraction.**

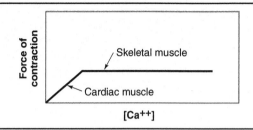

FIGURE 6.3

Figure 6.3 shows the force of contraction as it relates to Ca^{++} release from the SR. Notice the arrows that denote initial release of Ca^{++} from SR in response to action potential. As trigger Ca^{++} from the T-tubules comes in to cause the release of more Ca^{++} from the SR and have an effect itself, cardiac muscle reaches full (or greater) contraction strength.

d. **Effect of Autonomic Nervous System**

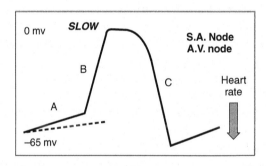

FIGURE 6.4

i. The <u>parasympathetic</u> **nervous system response** via M2 receptor stimulation. These receptors decrease HR via a G-protein mechanism.
ii. We can see that the slope of A has changed. We are now **decreasing the rate at which K^+ permeability progressively declines (causing prepotential to move toward 0 slower).**
iii. We may also hyperpolarize to begin with (-70 mV instead of -65 mV).
iv. <u>Sympathetic</u> **stimulation** increases **HR** and **the force of contraction.**
v. Heart rate originates in the pacemaker tissue (SA node). We now increase the rate at which K^+ permeability progressively declines to increase HR.

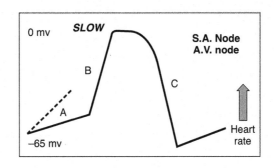

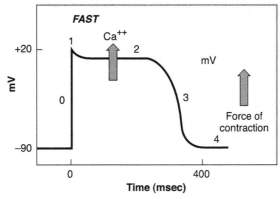

FIGURE 6.5

  vi. The force of contraction of cardiac muscle fibers is also affected: In response to sympathetic stimulation, the heart contracts more forcefully by increasing the amount of trigger calcium entering the muscle fibers.

e. **Gross Morphology: Heart is more on the left side of chest.**
  i. **Pericardial sac:** The heart is contained and held in place within this.
    (1) There are **two layers:** A tough layer called the **fibrous pericardium** (it anchors the heart and keeps it from over-extending), and an inner layer called the **serous pericardium,** which is very thin and delicate.
    (2) **Pericarditis:** inflammation of the pericardium
  ii. **Heart wall:** There are three layers.
    (1) **Epicardium:** On the outside. In the space between this and serous pericardium is the **pericardial cavity.** This cavity is filled with **pericardial fluid** to lubricate, as these layers constantly move past each other.
      a. **Cardiac tamponade** is a condition in which there is an excess of pericardial fluid. This increased pressure partially compresses the heart.
    (2) **Myocardium:** ventricular and atrial muscle fibers that are the bulk of the heart and provide the force
    (3) **Endocardium:** Part of the heart that comes in contact with the blood; it lines the inside of the heart and valves. Its role is in regulating cardiac function, and it is an important component of the anatomy of the heart.

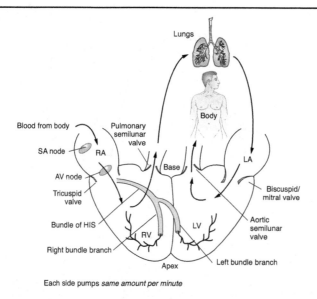

Each side pumps *same amount per minute*

FIGURE 6.6

iii. Chambers and Flow
  (1) The heart **is not heart-shaped** (really, the atria hang out on the side).
  (2) Blood is coming from the systemic circulation into the right atrium, through the tricuspid valve into the right ventricle, through the pulmonary semilunar valve to the pulmonary arteries, to the lungs, back to the left atrium through the pulmonary veins, through the bicuspid valve (mitral) to the left ventricle and through the aortic semilunar valves and into the aorta to the body (systemic circulation).
  (3) **Need a lot more cardiac muscle** to pump blood through the systemic circulation than through the pulmonary circulation
  (4) The heart consists of two functional syncytia:
    a. **Two atria:** Left and right are interconnected electrically so the action potential propagates to both quickly.
    b. **Two ventricles:** Action potential in the ventricles spreads through all the ventricles quickly.

III. **Mechanical Events of Heart**
  a. **Frequency of Depolarization and Velocity of Conduction**

ELECTRICAL ACTIVITY OF HEART

High

Frequency of depolarization	Velocity of conduction
– SA node	– Purkinje fibers
– AV node	– Muscle fibers
– Contractile elements	– AV node and bundle

Low

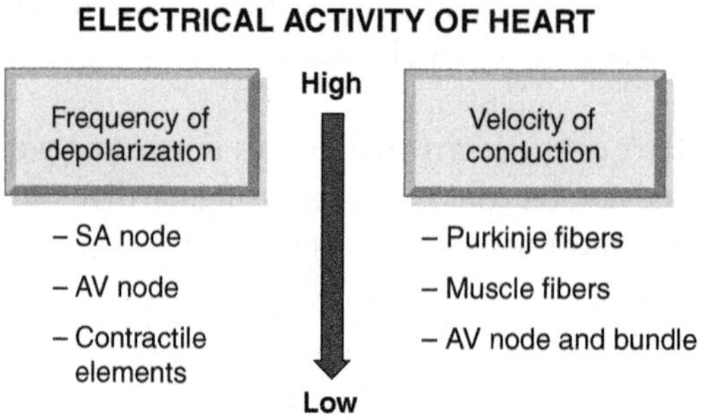

FIGURE 6.7

    i. The HR is dictated by the rate of the most frequently depolarized structure (SA node spontaneously depolarizes).

    ii. You do not want the atria and the ventricles to contract simultaneously, but they can relax simultaneously.

    iii. If you do not want the mechanical events to occur at the same time, then do not let the electrical events occur at same time.

    iv. The action potential goes to the AV node and the Bundle of His, but there is a decrease in velocity of transduction. Thus, **the action potential is delayed** before it gets to the bundle branches of the ventricles.

b. Terms

Traditionally, "systole" means "ventricular systole" and "diastole" means "ventricular diastole."

    i. Relaxation: diastole

    ii. Contraction: systole

    iii. Atrial and ventricular systole do not happen at the same time, but atrial and ventricular *diastole* can happen at the same time.

c. Cycle of Heart

    i. Figure 6.8 represents one cardiac cycle: The **outside area of the picture represents ventricular events and the inside area represents atrial events.**

      (1) At a rate of 72 beats per minute, a complete cycle occurs in 8/10 of a second.

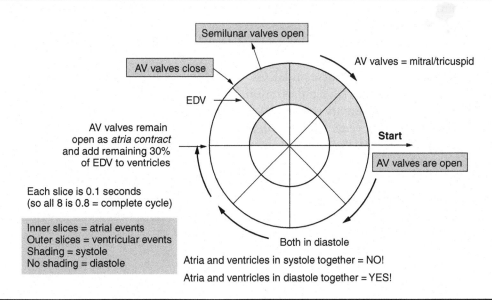

FIGURE 6.8

    ii. In the above picture the heart fills with blood during diastole (begin following around 6 o'clock on Figure 6.8) as the ventricles fill up to about 70% of end-diastolic volume (EDV).

    iii. Both are in diastole until the atria contract to fill the remaining 30% of the ventricles.

      (1) EDV: 120–130 ml of blood at the end of ventricular diastole (for *one* ventricle).

    iv. The ventricles then begin to contract. This causes the one way atrioventricular valves to quickly shut (heart sound number 1). The ventricles then undergo a definite period of **isovolumetric contraction,** when they are contracting, yet the volume is not changing.

When pressure has become great enough, the semilunar valves open and blood flows out of the ventricles. The amount of blood (70-90mL) ejected from *each* ventricle is referred to as the stroke volume (SV). At the end of ventricular systole the semilunar valves shut (second heart sound), and the atrioventricular valves once again will be pushed open to begin another cycle.

EDV − SV = end-systolic volume (ESV)

Ejection fraction = SV/EDV

What does sympathetic stimulation do to ESV? Ejection fraction?

If you slow HR, what happens to EDV?

If EDV were to increase, what does Starling's Law tell us?

IV. Electrical events (EKG)

a. William Einthoven described electrical events of the heart. The electrocardiogram (EKG) simply compares polarity of body surfaces in reference to each other. What you see on the EKG is electrical activity of muscular cells (in general, the electrical conducting cells are too small to pick up).
   i. This is *not* an exhaustive course on EKGs.
       (1) P waves represent atrial depolarization.
       (2) QRS complex represents ventricular depolarization.
       (3) T waves represent ventricular repolarization.
       (4) On an EKG strip, each small horizontal square is 0.04 seconds; each bigger square is 0.2 seconds.
       (5) Use the Dubin 300-150-100-75-60-50 method to quickly count/estimate HR **(see suggested readings).**

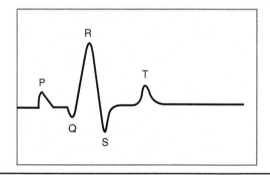

FIGURE 6.9

ii. What about atrial repolarization? It occurs at the same time as the QRS complex, so you cannot see it. The QRS complex has a greater magnitude and obliterates it.

iii. The P wave is an upward deflection in terms of the right arm in respect to the left leg.

iv. There are typically 12 leads:

   (1) Six limb leads detect electrical activity on anterior/lateral chest wall in the *frontal* plane.

   (2) Six chest leads detect electrical activity transversely through heart.

   (3) Leads I, II, and III *plus* AVR, AVL, and AVF are *frontal*.

   (4) V1–V6 are *transverse*.

   (5) *Focus on limb lead 2* for now:

      a. Imagine a triangle as shown in Figure 6.10. This is Einthoven's triangle (limb leads shown).

      b. Now imagine that the heart cells are positive *outside* relative to the inside. When these cells depolarize **not all at the same time** (e.g., from base to apex or from one side of atrium to the other), this change to negative on the outside of *some* cells is conducted as current. Then, a short time later, the signal reaches *all* cells of that depolarization event (e.g., both atrial muscle cells). At this point the signal is back to baseline because there is *no difference* between the cells at this point—they are *all depolarized*.

      c. Review: when there is a potential difference because one part of a cell group that has depolarized (it hasn't yet spread to all cells in that group), a deflection is made. This returns to baseline when all cells in that group are depolarized. When they then *repolarize* again, not all exactly at once, we see another deflection in signal.

      d. If we consider only lead II, then we can visualize these things (Figure 6.11).

      e. Consider that any time a signal *runs into* lead 2 it causes deflection *except* when it hits that lead in a perpendicular manner. Signal is maximum if it runs parallel to a lead!

      f. If the signal runs from negative to positive in reference to the lead, then an *upward* deflection is made.

      g. If the signal runs from positive to negative in reference to the lead, then a *downward* deflection is made.

      h. Explanation of events seen in lead II (Figure 6.11): Think, when the short, thick arrow goes from positive to negative along lead II, this is a negative deflection on the EKG. When lead II runs negative to positive, this is a positive deflection on the EKG. Also, when the arrow runs parallel to the lead, the signal is large on the EKG, and when the arrow is perpendicular to the lead, it is very small. **Understand these!**

         i. The ventricles depolarize from the septum to the *right* ventricle first. This vector is shown in Figure 6.11a (hollow arrow). Notice that it is close to perpendicular (so *almost* no signal detected = small deflection) and is running in lead II from positive to negative (which makes it a downward deflection). This is Q in the EKG.

         ii. The largest signal is depolarization from the endocardium to the epicardium throughout the ventricles. This shows up as the vector shown in Figure 6.11b (hollow arrow). Notice that it is close to parallel (so a large signal/deflection) and is running in lead II from negative to positive (which makes it an upward deflection). This is R in the EKG.

         iii. Finally, the ventricles depolarize from the apex to the base. This shows up as the vector shown in Figure 6.11c (hollow arrow). Notice that it is running in lead II from positive to negative (which makes it a downward deflection). This is S in the EKG.

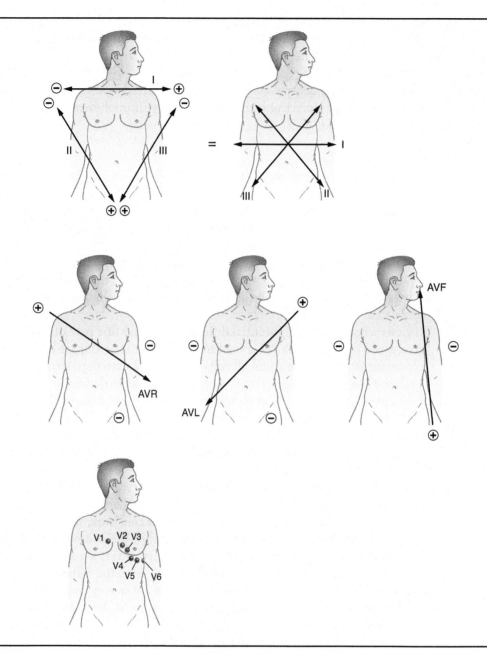

FIGURE 6.10

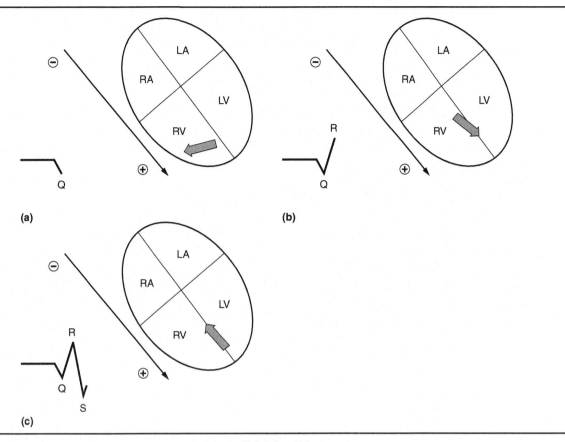

FIGURE 6.11

Answer *from base to apex* or *from apex to base* for each of the following:

How do ventricles depolarize?

How do ventricles repolarize?

Does the P wave represent ventricular systole?

Does the P wave represent atrial systole?
(Remember what the EKG tells you.)

V. **Abnormal Electrical Events**
  a. Abnormal rhythmicity of pacemaker
     i. **Tachycardia:** Rapid HR; at rest it is greater than 100 beats per minute.
        (1) The rhythm is normal; it is just too fast.

(2) Caused by an increase in temperature, acute hyperthyroidism, anxiety, or increased sympathetic effect.

(3) Pathological reason: **Heart muscle is weakened** and it cannot contract forcefully, so it beats more frequently.

   ii. **Bradycardia:** Slow HR; at rest it is less than 60 beats per minute.

(1) The rhythm is normal; it is just too slow.

(2) May be due to **carotid atherosclerosis**—stimulates baroreceptors due to narrowing of artery wall. Sends signals to decrease HR (to decrease BP).

(3) **Athletic bradycardia**: Affects athlete with strong heart (endurance athlete)—this is *not* pathological.

  b. **Abnormal rhythmicity resulting from blocks**
    i. **Atrioventricular (AV) blocks**: Most common.

(1) Causes: Due to coronary ischemia; scar tissue after inflammation (of AV node and bundle) blocks the impulse from the atria to the ventricles.

(2) **Types/Degrees:**
    a. **1st Degree heart block**
      Pattern is normal, but **the P-R interval is lengthened** (too much time between the atria and ventricular depolarization); blockage causes an increased time.

    b. **2nd Degree heart block**
      All QRS complexes are preceded by P waves, but not all the P waves are followed by QRS complexes. Depolarization wave is being blocked at the AV node (long time), and there is another P wave before QRS complex (the ratio of P waves to QRS complexes can be expressed as an integer—the ratio will stay the same, e.g., "3 to 2" or "4 to 2").

    c. **3rd Degree heart block**
      Complete heart block—complete dissociation between atrial and ventricular events; ***not*** expressed as an integer.

    ii. **Stokes-Adams Syndrome: Ventricular Escape**

(1) Atrioventricular (AV) block comes and goes in some patients.

(2) There are periods when that impulse is conducted and other periods when it is not.

(3) When atrioventricular (AV) conduction stops, ventricles do not pick up their own rhythm until a delay of several seconds; then the ventricles, once again, begin beating (ventricular escape) at a much slower rate (15–20 beats per minute).

(4) Since there is no perfusion of the brain and other tissues during this time, the person might faint.

  c. **Abnormal pathway of impulse transmission**
    i. **Accessory conduction pathway (Wolf-Parkinson-White syndrome):** This is a pathology where there is a **short circuit around the AV node and bundle.** The impulse is not delayed properly and reaches the ventricles prematurely.

    ii. **Circus movement and/or re-entry:** Anything that delays the electrical event across the heart can cause this. Delaying the signal allows it to "re-enter" and depolarize part of the heart again (erroneously).

(1) The electrical event enters where it is not supposed to, and a bizarre electrical event called a "circus movement" occurs.

(2) An **enlarged heart** is one way this can happen. It is not absolute refractory when the signal comes back around, so it depolarizes again.

(3) There can also be **damage to the conduction system** (perhaps from myocardial infarction) that causes the signal to take longer.

(4) **Drugs** may also decrease refractory time and lead to circus movements.

      iii. **Re-entry (fibrillation):** Multiple circus movements are occurring.
        (1) There are little clusters of muscle fibers all doing their own thing ("Anarchy of Events").
        (2) Ventricular fibrillation (V-Fib): "Anarchy" in ventricular muscle fibers because they are not controlled by SA node; this is life-threatening.
        (3) If someone is in V-Fib: **Try to depolarize everything so that the SA node will pick up normal rhythm and a "sinus" rhythm (use paddles).**
        (4) Atrial fibrillation (A-Fib): Ventricles fill 70% without atria, so this is not as big a deal (patient can live in A-Fib), but the blood does not go out of the atria and the patient may get blood clots. Such patients typically take an anticoagulant.
   d. Ectopic foci: Small area of the heart becomes hyperexcitable and depolarizes more frequently than the SA node. A new pacemaker causes premature depolarization.
      i. PACs: premature atrial contractions
      ii. PVCs: premature ventricular contractions, induced by stress, caffeine, lack of sleep, but indicative of cardiological myoischemia. The premature beat does not elicit a pulse pressure because it is so weak. Usually, these are not a big deal.

VI. Regulation of Cardiac Output

   a. Factors influencing: Cardiac output is the amount of blood the heart pumps in a minute (average male = 5,000 mL/min at rest, but much [five to seven times] greater during exercise).
      i. Two variables: **Cardiac output = HR $\times$ SV.** HR is the number of times the heart beats per minute, and SV is how much the heart ejects every time it beats, or the amount of blood the heart pumps each time the ventricles contract.
      ii. Cardiac Output Max - Cardiac Output Rest = Cardiac Reserve
   b. Intrinsic mechanisms: Exist only in the heart; involve stretching the atria or the ventricles
      i. Homeometric: when the atria are stretched and make the SA node more irritable and make it depolarize more frequently (increased HR). If we increase venous return, we increase pre-load (EDV), which stretches the atria, making the SA node more irritable (increased HR).
      ii. Heterometric (Starling's Law applied to the heart):
        (1) If you increase the stretch of the ventricles, then you increase the force of contraction by causing a more optimal alignment of thin and thick filaments. There is an optimal length beyond which contraction strength would decrease.
        (2) If you increase the filling of the heart (increased preload), then you stretch the ventricles; this would increase the force of contraction and increase the SV. Thus, it would increase the cardiac output.
      iii. What causes the ventricles or atria to stretch?

   c. Extrinsic mechanisms: Nervous system
      i. Parasympathetic: quick-acting (beat to beat)
        (1) Innervation: The left and right vagus nerve
          a. The left vagus nerve innervates the AV node and the atrial muscle fibers.
          b. The right vagus nerve innervates the SA node and the atrial muscle fibers.
          c. This can cause a decrease in HR by decreasing the rate at which K^+ permeability progressively declines during the prepotential (and slightly decreasing SV).
        (2) Works via G-proteins.
      ii. Sympathetic: slower-acting
        (1) Innervation: From the thoracic spinal nerves
          a. Increases the rate at which K^+ permeability progressively declines to increase the HR and increases SV by increasing the amount of trigger Ca^{++}
        (2) Effect and ionic explanation: works via cAMP mechanism

d. **Afterload:** pressure pushing back on the heart *as it contracts* (systolic BP). The heart must overcome the diastolic pressure and the systolic pressure in order to be contracting (**see Figure 6.12**).

e. **Preload:** amount of blood in the ventricles at the very end of ventricular diastole (EDV)

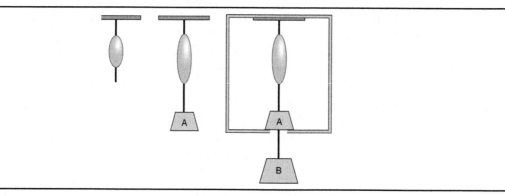

FIGURE 6.12

In Figure 6.12, we see three situations. On the left is a muscle that is hanging from a force transducer with no weight attached to it. In the middle is that same muscle stretched to its optimal length (Starling's law) with weight "A." On the right is the same muscle with "A" stretching it to an optimal length, but now it has a scaffold holding it just below "A" so that it cannot stretch any further than "A" makes it stretch. The scaffold is placed just below the point to which "A" stretches it and does not relieve any of the demand placed on it by "A." Weight "B" now is added below weight "A," and the scaffold prevents any further stretch. However, if the muscle were to contract, it would now need to overcome both "A" and "B" to shorten the muscle. This is one way of understanding the concepts of "preload" and "afterload," where "A" is preload and "B" is "afterload." **With regard to the whole heart, we would refer to EDV as the preload (stretching the heart; similar to "A" here) and to systolic BP as the afterload (the force against which the muscle contracts; similar to "B" here).**

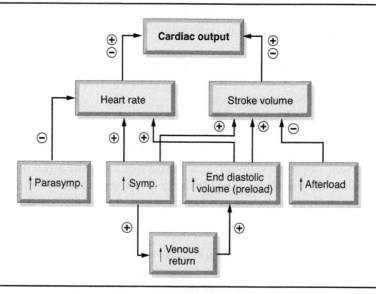

FIGURE 6.13

What affects cardiac output?

What would increasing afterload do to the force of contraction needed to pump blood into the aorta?

VII.   **Congestive Heart Failure:** There is some deficiency in the myocardium (degenerative); it causes the heart to not eject blood efficiently.
- a.   **Etiology:** There is an **increase in ESV**; the heart does not efficiently empty, causing an increase in ventricular pressure.
  - i.   The pressure backs up and increases the left atrial pressure, pulmonary pressure, and capillary pressure, eventually causing **pulmonary edema**.
    - (1) There is considerable mortality with pulmonary edema.
- b.   **Treatment**
  - i.   **Diuretics:** These simply decrease extracellular fluid to compensate for the edema.
  - ii.   **Cardiac glycosides:** These inhibit the mechanism that removes free Ca^{++} from the cardiac cell into the interstitial fluid (poison Na^+/K^+ pumps; more Na^+ in cell, so *less* Ca^{++} can exchange with Na^+ via Na/Ca exchange protein on membrane). Thus, more Ca^{++} is in cell. This increase or accumulation of Ca^{++} leads to a **more forceful contraction** and an increase in SV.
    - (1) Drugs: **digoxin (LANOXIN)**
      - a.   Note these have not led to a decrease in mortality; instead they seem to only relieve symptoms.

VIII.   **Coronary Blood Flow**
- a.   **Normal physiology:** The heart is supplied with blood—the myocardium is supplied with branches from the left and right coronary arteries, which come from the aorta beyond the left ventricle.
  - i.   Coronary blood flow is greater during diastole than during systole.
  - ii.   There is greater flow during diastole because blood has greater access to the heart during this period.
    - (1) **During systole, the semilunar valves are open and partially block the entry of the coronary arteries.** During diastole, the semilunar valves are closed and there is more access to the coronary arteries.
    - (2) When the heart contracts during systole, **it partially compresses the coronary arteries** (increases the resistance and decreases the flow into the coronary arteries).
  - iii. Blood flow to the heart is important because the heart has a tremendous demand for oxygen. It consumes more oxygen per gram than any other tissue in the body. When blood flows through the heart, 75% of the oxygen in the blood is removed at rest. Thus:
    - (1) **Blood flow to the heart is increased** when there is a demand for more oxygen.
    - (2) Must have coronary arteries with a sufficient diameter.
    - (3) If the demand is not met, then **myocardial ischemia** (reduced blood flow to the heart) occurs, or the blood flow to the heart is less than the demand of the heart for oxygen.
    - (4) If ischemia becomes more severe, it results in a heart attack (myocardial infarction).

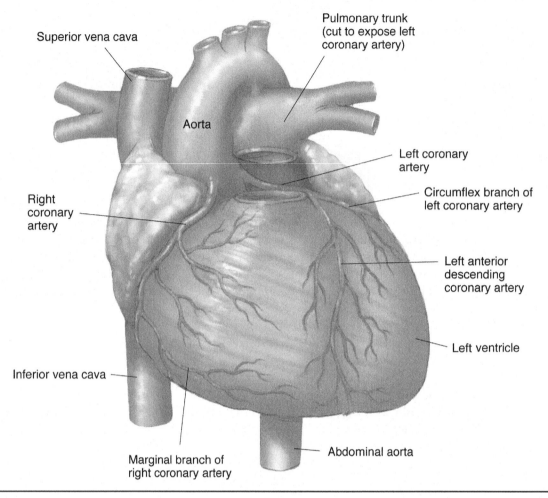

Superior vena cava

Pulmonary trunk
(cut to expose left
coronary artery)

Aorta

Left coronary
artery

Right
coronary
artery

Circumflex branch of
left coronary artery

Left anterior
descending
coronary artery

Left ventricle

Inferior vena cava

Marginal branch of
right coronary artery

Abdominal aorta

FIGURE 6.14

b. **Pathophysiology:** What causes myocardial ischemia?
  i. **Vascular spasm:** Spastic contractions of arterioles (which are serious but short-lived); typically, the patient survives.
    (1) **NO:** endothelial relaxing factor that keeps the arterioles from having vascular spasms
  ii. **Coronary atherosclerosis:** progressive degeneration of the arteriole wall (narrowing due to plaque buildup), which ultimately leads to a decrease in blood flow
    (1) **Mechanism:** an accumulation of lipids (rich in cholesterol), which are deposited in the endothelium lining the blood vessels. This can lead to a **"fatty streak"** in the arteries. The smooth muscle cells in the arteriole wall then migrate to the muscle layer, where they accumulate and form "plaque" buildup.
       a. **75% of college men** have fatty streaks in arteries.
    (2) **Risk factors:** high BP, high-cholesterol diets, diabetes, genetics, cigarette smoking, obesity, lack of exercise
    (3) Ischemia can lead to visceral pain called **angina pectoris**; occurs when the heart's demand for oxygen is greater than can be supplied
       a. **Can have ischemia without ever having angina.**
    (4) **Myocardial infarction** (i.e., heart attack): irreversible damage to myocardium

iii. **Thromboembolism:** a consequence of atherosclerosis
  (1) A piece of plaque breaks off and causes some bleeding, which eventually clots. The clot then breaks off and flows downstream to smaller vessels, which can block the blood flow, causing a heart attack if it lodges in a coronary artery.
  c. **Pharmacology**
  i. **Nitrates**
    (1) Nitroglycerine (**NITROSTAT, NITRO-DUR**): donates NO, which causes vasodilation and ultimately decreases preload and afterload, which decreases the work of the heart
      a. Administered in the form of a **patch or placed under the tongue;** not swallowed because the hepatic portal system would take it to the liver, where NO would be destroyed. Instead, we avoid the hepatic portal system by using nonhepatic venous return to the heart.
  ii. **Beta blockers:** work by decreasing HR and thus decreasing SV; thus, the oxygen demand on the heart is lowered
  iii. **Ca^{++} channel blockers**
    (1) **Mode of action:** Relieves the symptoms of angina by blocking the slow Ca^{++} channels. This decreases HR and the amount of trigger Ca^{++}, and thus the force of contraction also decreases, which lessens demand for oxygen by the heart.
    (2) (**ISOPTIN**)
    (3) (**PROCARDIA**)
    (4) (**CARDIZEM**)
    (5) (**NORVASC**)
    (6) (**ADALAT**)
    These may also be used for hypertension.

IX. **Valvular Diseases & Murmurs:** abnormal heart sounds; generally associated with cardiac disease; can have a functional murmur (without disease)

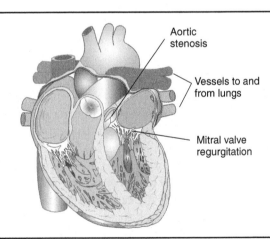

Aortic stenosis

Vessels to and from lungs

Mitral valve regurgitation

FIGURE 6.15

  a. **Examples**
  i. **Regurgitation (insufficiency, incompetency):** an abnormal sound from **the valve not closing properly—a whistle sound results**

    ii. **Stenosis:** murmur occurs when the **valve does not open completely** and the blood moves at a greater velocity; **makes a swishing sound.**

    iii. **Mitral valve prolapse:** flaps of the valve extend back into the left atrium when the left ventricle contracts, causing leaking; common in 5% of population but usually asymptomatic

  b. **Diagnosis:** usually can hear these things via stethoscope and then diagnose with an ultrasound (echocardiogram) of the heart

  c. **Causes**

    i. Rheumatic fever

    ii. Prolapse

X. Work of Heart: a function of **potential energy** (the development of high pressure, taking the blood from low pressure to high pressure) and **kinetic energy** (movement of blood)

  a. The work involved in potential energy is greatest (90%).

    i. **PE = BP × volume**

    ii. **KE = $(mv^2)/(2g)$** (this is 10%)

  b. Harder work hypertrophies the heart (muscle increases in size).

    i. As the muscle gets bigger, the vascularization cannot keep up—this can cause circus movements and other pathologies. **Chronic hypertension** places an extreme demand on the heart and leads to this hypertrophy.

    ii. **In athletes,** the heart gets bigger but the vascularation keeps up with the growth of the heart. The heart does not get nearly as large as in chronic hypertension.

    iii. For pressure in chambers like this, use Laplace's law: Tension in wall = (Pressure × radius of lumen)/(2 × thickness). We will revisit this with pulmonary physiology/ventilation.

      (1) **P = (2T)/R** *or* **T = (PR)/(2 × t)**

        (This formula explains why a bike tire with a small radius can withstand higher pressure, for the same wall tension as a larger car tire.)

        T = tension in wall

        R = radius/lumen of chamber

        P = pressure inside tube

        t = thickness of wall of chamber

        (See Figure 6.16.)

      (2) As the radius gets smaller, the pressure increases for any given tension.

    iv. Use Laplace's law to understand what happens **in hypertrophy of the heart:**

      (1) If systemic pressure increases, then P increases. Thus, the thickness of the wall must increase to maintain appropriate pressure. This is typical left ventricular hypertrophy in response to chronically high BP. This is concentric hypertrophy.

      (2) If there is a volume overload, then the radius of the chamber increases. In this case, there is hypertrophy to bring the heart back to a normal size. In this case, there is some hypertrophy of the wall *as well as* some thickening to maintain the appropriate tension. This is eccentric hypertrophy of the heart.

      (3) Note that in both cases, heart failure can result when there is chronic stress and the blood supply does not keep up with the muscle growth. While it is a bit more complicated than this, a failing heart from chronic stress eventually "fails" and dilates. (Note that an endurance athlete's heart enlargement is more than compensated for by an increase in many things, including filling time, contractility, oxygen extraction, etc.)

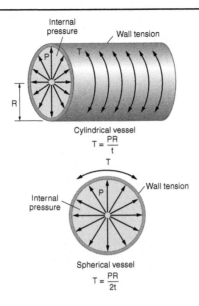

FIGURE 6.16

T = wall tension; P = pressure within chamber; R = radius of chamber, t = thickness of wall of chamber

SUGGESTED READINGS

1. Dubin, D. 2000. *Rapid interpretation of EKG's*, 6th ed. Tampa: Cover Publishing.
2. Guyton, A.C., and Hall, J.E. 2011. Chapters 9–13 and 20–22 in *Textbook of Medical Physiology*, 12th ed. Baltimore: Saunders.
3. Koeppen, B.M. and Stanton, B.A. 2008. Chapters 13-19 in *Berne & Levy Physiology*, 6th ed. St. Louis: Mosby.

CHAPTER 7

Cardiovascular Physiology II: Vasculature and Regulation

I. Introduction

   a. **Anatomic Overview:** Arteries, arterioles, capillaries, venules, veins
   b. Amount of Blood: There is more blood in the vasculature on the systemic side. **More blood is also in the systemic veins than in the systemic arteries.**
   c. Cross-Sectional Area: **Capillaries have the most total cross-sectional area**—there are billions of them!
   d. **Rate of Flow:** The velocity of flow is slowest in the capillaries.
   e. **Pressure Differences:** Complete Figure 7.1 during class as you discuss it.

FIGURE 7.1

II. Arteries

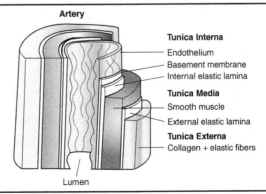

FIGURE 7.2

a. **Tunicas:** walls (three layers)
   i. **Interna:** composed of endothelium; touches the blood itself and has an internal elastic membrane
   ii. **Media:** smooth muscle; has an external elastic membrane
   iii. **Externa:** collagen and elastic tissue
b. **Function ("elastic recoil"):** There is continuous flow through capillaries while the heart beats and contracts—flow is continuous because of the elasticity of the arteries.
   i. **Elastic recoil:** When the heart goes into systole, it forces the blood down the arteries and stretches them.
   ii. **During diastole there is still pressure** because of the elastic recoil of the arteries (which also helps to propel blood).
c. **Pressure curve: Pulse pressure** is the difference between the systolic pressure and diastolic pressure.
   i. **Mean pressure:** the point on the curve where half the area is above the curve and half is below the curve
      (1) **Mean pressure is closer to diastolic pressure than systolic pressure because the heart is in diastole longer.**

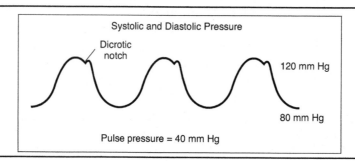

FIGURE 7.3

$$\text{MAP} = \text{DBP} + [0.333*(\text{SBP} - \text{DBP})] \; or \; \text{MAP} = 0.333*(\text{SBP}) + 0.666*(\text{DBP})$$

d. **Pathologies**
   i. **Atherosclerosis:** deposition of lipids in the smooth muscle layer and subsequent buildup of "plaque" (narrowing the lumen)
      (1) Greater the diameter, the greater the rate of flow. This is critical in areas such as coronary carotid and renal arteries.
   ii. **Aneurysm: bulging out of a vessel wall**
      (1) Often a consequence of atherosclerosis. If it bursts (**hemorrhage**), patient could bleed to death.
      (2) Most common location is the **dorsal aorta (abdomen)**; next most common is in the **cerebral vasculature.** The latter is obviously much harder to repair.

III. **Arterioles**
   a. **Anatomy:** small arteries that deliver blood to the capillaries. These have the same tunica layers as the arteries at the arteriole end, but not at the capillary end.
      i. There is a **resistance to flow in the arterioles—pressure falls quickly.**

b. Function: regulate flow to various organ systems (e.g., during exercise)

c. Regulation

   i. Autonomic nervous system: Arterioles supplied with alpha and beta receptors.

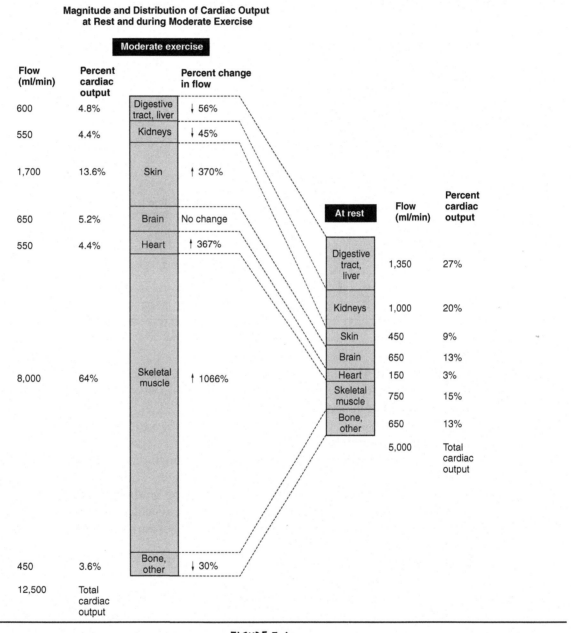

FIGURE 7.4

(1) The lengths of the sections in the above figure represent differences in cardiac output.

(2) During exercise, cardiac output must increase tremendously.

(3) Distribution of blood flow to different body parts is changed by **vasodilating in some areas and vasoconstricting in other areas.**
ii. **Local "regulators"** also regulate the distribution of blood flow to different places.
    (1) CO_2, H^+, O_2, prostaglandins, kinins, histamines, nitric oxide (NO), adenosine

IV. Capillaries: Almost all cells are intimately associated with capillaries.

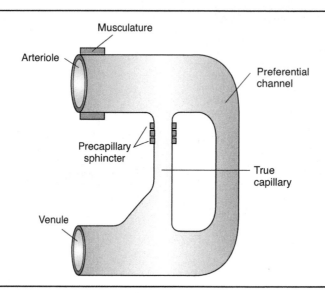

Musculature

Arteriole

Preferential channel

Precapillary sphincter

True capillary

Venule

FIGURE 7.5

a. **Anatomy:** There are 10–40 billion capillaries in the body. Almost no cell is more than 1/100 of a centimeter from a capillary (exceptions are epidermis, cornea of eye, and cartilage). The diameter is such that an RBC can barely get through.
   i. Capillaries are microscopic vessels that run from arterioles to venules.
      (1) They contain a **single layer of endothelium.**
      (2) They have slits between adjacent cells that are so small that a protein or bigger cannot go through.
      (3) They can be placed into three categories, depending on their "leakiness."
          a. Continuous: most common. This type allows molecules up to the size of proteins to migrate from plasma to the interstitial fluid. These are not found at the blood-brain barrier.
          b. Larger: found in SI, kidney, and exocrine glands
          c. Large gaps: found in liver sinusoids
   ii. Capillaries are in a circuit. Blood comes into an arteriole and then can get to the venules by going around the tissue (**preferential channel**), or it can go through the tissue via a **true capillary**; the capillary bed is where the nutrient and gas exchange takes place.
      (1) Capillaries are so small that RBCs usually will distort to get through.
      (2) The **precapillary sphincter (smooth muscle)** of the true capillary decides which direction the blood goes.
      (3) The sphincter **is not** innervated by the autonomic nervous system.
          a. If flow were regulated by the autonomic nervous system (ANS), then you would have ~40 billion capillaries to try to regulate.

(4) **Sphincters**: If they are closed, then blood goes the fast route; if they are open, then blood flows through the capillary bed.

(5) **Local control is ideal**.

iii. **There are two theories** of local control:

(1) **Vasodilator theory**: Buildup of some vasodilating substance (carbon dioxide or hydrogen ion) causes the sphincters to relax and allows blood to flow through.

(2) **Oxygen demand**: If there is plenty of oxygen, then the sphincters close. If the oxygen is low (used up) then the sphincters relax and the blood flows through the bed.

(3) This is called **"regulation of microcirculation."**

b. **Exchange Between Fluid Compartments**

i. **Diffusion**: As stuff flows through the capillary, nutritional demands of the tissue are met via diffusion. Diffusion depends on the concentration gradient, solubility, and whether or not that substance is going to pass through the epithelial cell itself or through the slits.

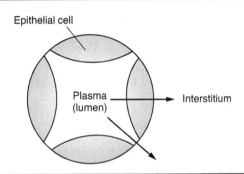

FIGURE 7.6

ii. **Filtration (Bulk Flow)**: the net movement of fluid through some semipermeable membrane in response to a pressure gradient. As blood comes through the capillaries (from the arteriole end to the venous end, there is always a net loss of fluid from the plasma to the interstitial fluid (less volume at the venous end).

(1) **Pressure at the Capillary Level**

Arterial ——————————→ Venule		
BP	+30	+10
OPPP	-28	-28
IFP	+3	+3
IFCOP	+8	+8
Net	+13	-7

So + 6 mm Hg is pressure driving fluid *into* interstitium.

BP: blood pressure
IFP: interstitial fluid pressure (kind of a sucking into the interstitium, since the interstitial fluid drains via the lymph, a one-way drainage system)
IFCOP: interstitial fluid compartment osmotic pressure
OPPP: osmotic pressure of plasma proteins

    a. Filtration of fluid from the capillary to the interstitial compartment is indicated by (+).

    b. Filtration of fluid from the interstitial fluid to the capillary is indicated by (-).

(2) The net effect of all the pressures is a loss of fluid into the interstitial fluid; too much fluid in the interstitial fluid causes edema.

(3) **The lymphatic ducts** drain fluid from the interstitial space and dump the fluid back into the vascular system.

(4) **Pathologies/Applications:**

a. **Pregnant women:** edema in the lower extremities during late pregnancy because the presence of the fetus impairs venous return; pressure in the capillaries goes up and more fluid is filtered out into the interstitium than can be taken up by the lymph

b. **Elephantiasis:** lymph nodes are blocked and lymphatic return is decreased; scrotum and lower extremities are common areas

c. **Congestive heart failure:** causes **pulmonary edema.** As the heart loses the capacity to eject blood, SV decreases and ESV increases. Ventricular and atrial pressures increase, fluid builds up in the pulmonary capillaries, and the fluid pushes out into the pulmonary interstitium.

d. **Cut finger:** inflammation causes the capillaries to become more permeable

e. **Starvation:** muscles are broken down because the body must begin "digesting" plasma proteins (to make up for lack of protein in diet); this causes OPPP to decrease, and fluid oozes from the liver into the abdominal cavity **(ascites).**

f. **In women who have mastectomy: Why is there the possibility of edema?**

g. **Liver failure: Ascites—fluid flows into the abdominal cavity.**

V. Venules and Veins

a. **Anatomy:** Valves—**One-way valves** promote venous return to the heart.

i. As veins get closer to the heart, the total cross-section of veins decreases and the velocity increases.

ii. These vessels are **distensible capacitance** vessels—there is not much elastic recoil; they are able to stretch to accommodate increases in volume, but there is not a considerable increase in pressure.

b. **Functions:** There is a huge storage of blood volume in veins.

i. The sympathetic response decreases the amount of blood in the venous system and diverts it to skeletal muscle.

ii. There is **not a big pressure gradient in the veins** (~15 mm Hg in veins to ~0 mm Hg in the heart); normal posture is standing, so veins are located between skeletal muscle masses:

(1) The contracting muscles squeeze the veins so that the blood can flow back to the heart **(the "muscle pump").**

(2) Valves in the vessels keep the blood moving in the right direction.

iii. During cardiac bypass surgery, a graft is made out of a vein from the leg; **the new vein must be grafted in the correct direction!**

c. **Varicose Veins & Hemorrhoids**

i. **Varicose veins:** Blood pools instead of continuing toward the heart. Valves become dysfunctional.

ii. Veins dilate and engorge (usually in the superficial veins of the leg).

iii. Varicose veins are more common in women who have been pregnant.

iv. **Hemorrhoids:** varicose veins in the rectum

d. **Phlebitis:** an inflammation of veins

VI. Hemodynamics
   a. **Flow**
      i. $Q = \dfrac{\Delta P}{R}$
         (1) Q = flow
         (2) ΔP = Δ pressure
         (3) R = resistance
      ii. **Laminar:** the blood nearest the walls travels a bit slower than the middle of the blood flow; this causes the blood flow to have a smooth, bullet-shaped flow, known as **laminar flow.**
         (1) **Laminar flow is very quiet.**
      iii. Indirect blood pressure—To take BP:
         (1) Consider a normal BP as ~120/80 mm Hg.
         (2) Inflate the cuff by pumping until the gauge is about 140–160 mm Hg, then release the cuff pressure slowly.
         (3) The first noise will be heard as the cuff pressure decreases to around ~120 mm Hg and the pressure inside the vessel is great enough to push blood through. **During systole, the pressure in the arteries is greater than the cuff pressure and thus you will hear resultant turbulent flow.**
         (4) As the cuff pressure drops below ~80, all sounds will disappear as laminar flow returns. The sound disappears because the arterial pressure is now always greater than the cuff pressure. **During diastole, the cuff pressure has become less than the pressure of the arteries when relaxed (laminar flow returns).**
   b. **Pressure:** BP = cardiac output × total peripheral resistance. You can increase pressure two different ways:
      i. **The hard way:** Increase the cardiac output, which is a function of SV and HR.
      ii. **The easy way:** Increase the resistance.
         (Remember, standard temperature and pressure [STP] = 0°C and 760 mm Hg.)
         We know from introductory chemistry and physics that things flow from high pressure to low pressure. If normal BP is 120/80 mm Hg, and atmospheric pressure is 760 mm Hg, then when your arm gets sliced open, why doesn't air rush into your vessels instead of blood rushing out like it does?
   c. **Resistance**
      i. $TPR = \dfrac{8\eta l}{\pi r^4}$
         (1) η ("eta") = the viscosity of blood (the greater the viscosity of the liquid, the greater the resistance to flow); RBCs give the blood the greatest viscosity.
         (2) l = length of vessel; the longer the vessel, the more resistance to flow.
         (3) r = radius of the vessel; as the radius gets smaller, there is an increase in resistance.
         (4) TPR = total peripheral resistance
      ii. **Poiseulle's law:**
         $$Flow = \dfrac{(\pi)\, \Delta P\, (r^4)}{8\eta l}$$
         (1) Flow is a function of the radius and change in pressure.
         (2) If you double the viscosity, the magnitude of the flow decreases by half.
         (3) The radius changes if you want to increase the flow by vasodilation or vasoconstriction (predominantly in the arterioles).
         (4) **r: You don't have to change this much to make a difference.**

VII. Regulation of Blood Pressure

a. **Neuronal mechanisms:** To regulate the arterial BP: adjust the cardiac output, the total peripheral resistance, SV, HR, or the radius.

   i. The heart is the predominant structure for SV and HR. The arterioles are the predominant structure for total peripheral resistance and radius.

   ii. In the neuronal regulation of BP, there is a visceral reflex with afferent and efferent pathways, an integrating circuit, and an effector.

   (1) **The integrating circuit is in the brain** (medulla regulates vital functions).

   iii. The medulla houses the **nucleus of the tractus solitarius (NTS). The NTS will affect ventilation, but for now let's focus on the role it plays in neuronal controls of BP.** The NTS, when stimulated, affects many other medullary centers:

   (1) **VMC: vasomotor center** (sympathetic). This center may send signals via the T1–L3 spinal nerves to α-1 receptors on arterioles and N-1 receptors on the adrenal medulla. This increases BP.

   (2) **VC/CIC: vagal center/cardio-inhibitory center** (parasympathetic). This center sends signals via the vagus nerve to the M2 receptors on the heart to decrease HR and thus decrease BP.

   (3) **CAC: cardioaccelleratory center** (sympathetic). This center sends signals via T1–T4 spinal nerves to increase HR and contractility of the heart.

   (4) <u>**At this point it would be *very* helpful to see Chapter 10, *Ventilation*, for an introduction to the terms and a more detailed and accurate description of neuronal regulation.**</u>

   iv. Effectors in the reflex: arterioles and the heart (myocardium)

   v. **Parasympathetic efferent neurons:** The parasympathetic pathway between the vagal center and the heart is a cranial nerve (vagus nerve).

   (1) There is a long preganglionic parasympathetic neuron; here (the synapse at the ganglion between the two neurons) the neurotransmitter is ACH, with N1 receptors.

   (2) If you stimulate the N1 receptors, then HR decreases.

   vi. **Sympathetic efferent neurons:** Remember that **cranial nerves do not house sympathetic function.** These must leave from a spinal nerve.

   (1) Long, postganglionic sympathetic neurons with the neurotransmitters norepinephrine (~95%) and epinephrine (~5%).

   (2) Increase HR (β-1) so the K^+ permeability declines quicker in the prepotential and there is an increased force of contraction. This increase in SV is because more Ca^{++} is in the cell.

   vii. **Innervation of arterioles:** There is no parasympathetic innervation to the arterioles (except to arterioles that serve erectile tissues); there are muscarinic receptors (M2) on all arterioles that are not innervated.

   (1) Sympathetic: Long, postganglionic, with alpha-1 and beta-2 receptors.

   a. **Alpha-1 = vasoconstrict; beta-2 = vasodilate** (on arterioles)

   (2) Blood vessels supplying both the heart and certain skeletal muscles have more beta-2 receptors and thus there is greater blood flow to these areas during times of sympathetic duress. Note that during whole-body distress, skeletal muscle beds must be constricted to some degree or syncope will result.

   viii. **Afferent Pathways**

   (1) **Baroreceptors:** There are receptors at the **bifurcation of the carotid arteries** and at the **arch of the aorta,** which are sensitive to being stretched; increased BP stretches the baroreceptors. If there is increased stretch, this increases the number of signals

sent via the afferent neuron. Afferent pathway is **the vagus nerve (X) from the arch of the aorta** and **the glossopharyngeal nerve (IX) from the bifurcation of the carotid arteries.**

    a. Very active around physiological BP

    b. The visceral reflex running in the 9th and 10th cranial nerves; this is the afferent limb—it runs in the parasympathetic pathway.

    c. **Overall: Increased baroreceptor signals increase NTS signals to inhibit the VMC/CAC and excite the CIC/VC.**

(2) Chemoreceptors: At the bifurcation of the carotid arteries (cranial nerve IX) and the arch of the aorta (cranial nerve X).

    a. Sensitive to low-oxygen, high-hydrogen ions and/or elevated CO_2

    b. When they are stimulated, they send signals to the NTS to stimulate the VMC/CAC and inhibit the CIC/VC (think opposite of baroreceptors—these want to increase BP).

    c. We do not use these very often; they do not become active toward PO_2 until it drops to around 60 mm Hg.

    What happens to arterial PO_2 when you go to altitude (at least 5,000 ft)?

    What happens with your peripheral chemoreceptors?

(3) CNS ischemic receptors: On the medulla (of the brain) there is a chemosensitive area—sensitive to high CO_2 (really a high $[H^+]$)—that when excited has a powerful effect on the VMC. The main effect here is the sensitivity to the H^+s. **In fact, this stimulation, as we will see in the respiration section, is what drives respiration.**

b. "Chemical" Mechanisms

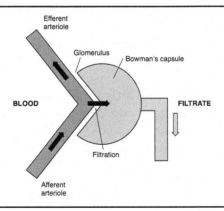

FIGURE 7.7

Figure 7.7 shows the interface between the renal arterial blood and the tubular system of the kidney. Figure 7.8 shows the conversion of angiotensinogen to angiotensin II. As we will see, angiotensin II does many things.

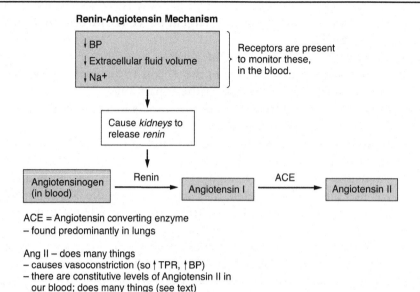

Renin-Angiotensin Mechanism

↓ BP
↓ Extracellular fluid volume
↓ Na+

Receptors are present to monitor these, in the blood.

Cause *kidneys* to release *renin*

Angiotensinogen (in blood) —Renin→ Angiotensin I —ACE→ Angiotensin II

ACE = Angiotensin converting enzyme
– found predominantly in lungs

Ang II – does many things
– causes vasoconstriction (so ↑ TPR, ↑ BP)
– there are constitutive levels of Angiotensin II in
  our blood; does many things (see text)

FIGURE 7.8

i. **Renin-Angiotensin:** Filtration occurs into Bowman's capsule.
   (1) The amount filtered has to be kept constant; known as GFR (glomerular filtration rate).
   (2) It is important for the GFR to remain constant in the cleansing of blood; that is why dialysis is required every couple of days for patients with kidney failure.
   (3) **There are mechanisms to monitor GFR:**
      a. **JGA (juxtaglomerular apparatus)** is the quality control device. The JGA monitors GFR by monitoring filtrate concentration.
      b. Filtration rate is affected by the constricting or dilating of the afferent and efferent arterioles.
   (4) If you increase the glomerular blood pressure (GBP), then you increase the GFR.
   (5) When the JGA senses the GFR is too low, it releases the hormone renin. Renin catalyzes the conversion of angiotensinogen (a plasma protein in the blood) into angiotensin I, which is converted into angiotensin II by the enzyme ACE (angiotensin converting enzyme).
   (6) **ACE is produced predominantly in the lungs.**
   (7) Angiotensin II is a vasoconstrictor. Along with the alpha and beta receptors on the arterioles, there are also angiotensin II receptors. These cause vasoconstriction, especially on the efferent arterioles of the kidney (causing increased GFR).
ii. **Antidiuretic hormone (ADH, vasopressin)** works at the kidney to reabsorb water back into the blood.
   (1) Was first known as "vasopressin" because it also causes vasoconstriction if in larger amounts. **The two main stimuli for ADH release are increased osmolarity of blood and angiotensin II.**
   (2) If one is hypohydrated, the osmoreceptors in the hypothalamus are stimulated. This causes the posterior pituitary to release ADH into the blood, which mainly acts to decrease water loss at the kidney level. It can also cause vasoconstriction if present in larger doses (if one were to go into hypovolemic shock). **Usually a blood volume loss of more than ~10% is needed as a stimulus to release ADH.**

VIII. Hypertension

a. **Introduction:** A disease of the circulatory system characterized by persistently elevated systolic or diastolic BP.

   i. 20–30 million Americans are affected.

   (1) Mortality due to hypertension = 1 million deaths per year.

   ii. **Diastolic BP: >90 mm Hg** is hypertensive, regardless of systolic BP.

   iii. **Systolic BP: >140 mm Hg** is hypertensive, regardless of diastolic BP.

   iv. Normal standards are attained with a BP cuff at the doctor's office (at rest).

b. **Types**

   i. **Primary ("essential hypertension"):** etiology unknown

   (1) About 85% of hypertensive patients have this type of hypertension.

   (2) Treat symptomatically, since etiology is unknown.

   ii. **Secondary:** etiology is known (~15% of hypertensive patients)

   (1) May be **renal hypertension** due to atherosclerosis; renal artery is clogged, so blood flow decreases, GFR decreases, renin secretion increases, and angiotensin II synthesis increases, causing hypertension.

   (2) May be due to **pheochromocytoma:** tumor of the adrenal medulla that causes oversecretion of epinephrine

c. **Treatment**

   i. **Nonpharmacological:** weight loss, change in diet, exercise, less stressful lifestyle

   (1) Best to control this and other things with diet and exercise before it happens!

   ii. **Pharmacological:** usually a **stepwise treatment**; try to use the minimum to get the maximum effect.

   (1) **Most will prescribe a diuretic first. If something stronger is warranted, then move on to using such things as beta-blockers, angiotensin-converting enzyme (ACE) inhibitors, and Ca^{++} blockers. An α-1 blocker could be used, but more as a last resort because of the resultant orthostatic hypotension.**

IX. Angiotensin II Pharmacology

a. **ACE inhibitors: (VASOTEC, ZESTRIL, ALTACE)**

b. **Angiotensin II receptor blocker: Angiotensin II is made, but it cannot bind its receptor on arterioles (COZAAR, DIOVAN).**

How would you treat a patient with hypertension who did not respond to a diuretic?

How would you treat a hypertensive patient who also has congestive heart failure?

X. Circulatory Shock

a. **Introduction:** Circulatory shock occurs for two reasons.

   i. Heart could be problem: Weakened heart causes decrease in BP.

   ii. A disproportionality between blood volume (too little blood) and the capacity of the vasculature (too much vasodilation)

   iii. **The bottom line in shock:**

   (1) The capacity is too big (vasculature) for the blood being pumped.

(2) There is inadequate blood flow to the organs, and the tissues cannot be maintained; **death will quickly result from circulatory shock if there is no intervention!** (It is a nasty feed-forward loop—the heart receives less venous return and thus pumps less blood back out, the heart pumps in a weaker manner because EDV is decreased, and vital tissues receive less and less perfusion.)

    iv. Hypotension **but *not shock***
- **(1) Orthostatic (postural) hypotension:** due to posture; if patient is lying down and suddenly rises to feet, there is not enough blood flow to the brain.
  - **a.** This happens to people who have been bedridden for awhile.
  - **b.** This also happens to people who are on certain medications (e.g., α-1 antagonists!).
- **(2)** Clinical disorder: **micturition vasovagal syncope** (vagus nerve fainting after urination)
  - **a.** Sleeping is parasympathetic and urination is parasympathetic: causes the BP to fall sufficiently so that you faint. Happens to many men in the middle of the night—too much parasympathetic effect.
- **(3) Emotional phenomenon:** The limbic system also influences the VMC.
  - **a.** Strange phenomenon—if one person faints, another usually will also faint. Also, some people have needle aversions, whereas others may faint in response to offensive language. These are all indicative that conscious brain centers modulate fainting in some way.

b. Causes of shock
- **i. Hypovolemic**: not enough blood volume; may be due to severe hemorrhage, profuse sweating, or severe diarrhea.
- **ii. Cardiogenic: most common type of shock**
  - **(1) Problem is the heart**—it is weakened and cannot adequately pump blood.
- **iii. Vasogenic: problem is the vasculature** (widespread vasodilation)
  - **(1) Septic: next most common type of shock.** Also known as blood poisoning (if you get a splinter and an infection ensues, it swells up and you get edema because of widespread vasodilation).
    - **a.** In septic shock, a toxin spreads to the whole body and widespread vasodilation occurs.
    - **b.** Results from periphery damage, appendix rupture, stab wound, or instrumental abortion; all can cause widespread vasodilation and decrease BP.
  - **(2) Anaphylactic:** due to antigen-antibody response. When you breathe in something in the air you are allergic to, histamines are released and you get vasodilation of arterioles in pharyngeal areas. This is a similar problem—but it is widespread. May result from **bee stings, horse serum, or food allergies.**
    - **a.** Phenomenon happens systemically; **widespread histamine production and widespread vasodilation occur.**
- **iv. Neurogenic:** Traumatic pain can cause shock.
  - **(1)** Higher centers of the brain influence the vagal center and the vasomotor center, causing shock due to intense pain.

c. Treatment: Give pain medications if needed.
- **i. Main goal is to restore pressure.**
  - **(1)** May need to give blood plasma.
- **ii. Drugs (DEXTRAN)**—a large, inert molecule that will not go through the capillary walls but will exert osmotic pressure; a "plasma expander" that causes the fluid from the interstitial fluid to be pulled into the capillaries/plasma so that the pressure may increase. **Why can we not just infuse these shock patients with water?**

SUGGESTED READINGS

1. Guyton, A.C., and Hall, J.E. 2011. Chapters 14–19 and 24 in *Textbook of Medical Physiology*, 12th ed. Baltimore: Saunders.
2. Koeppen, B.M. and Stanton, B.A. 2008. Chapters 15-19 in *Berne & Levy Physiology*, 6th ed. St. Louis: Mosby.

CHAPTER 6 AND 7 QUESTIONS

1. _____ Cardiac muscle
   a. Is striated
   b. Contains a well developed T-tubular network
   c. Can be paralyzed with curare
   d. All of these
   e. Two of these

2. _____ During the prepotential
   a. There is a progressive increase in potassium permeability
   b. There is a decrease in sodium permeability
   c. Calcium is moving out of the cell
   d. None of these
   e. Two of these

3. _____ During the plateau phase of the action potential
   a. Trigger calcium is released from the SR
   b. Calcium channels are open
   c. The membrane potential is a positive number
   d. All of these
   e. Two of these

4. _____ The first heart sound
   a. Occurs during ventricular systole
   b. Occurs during atrial diastole
   c. Is due to the closing of the AV valves
   d. All of these
   e. Two of these

5. _____ If a person's BP was 144/84
   a. The person would be prehypertensive
   b. The pulse pressure would be 60
   c. The mean pressure would be 114
   d. All of these
   e. Two of these

For questions 6–16, choose one of the following:
   A. Increases or is greater than
   B. Decreases or is less than
   C. Has no effect or is equal to

6. _____ The number of P waves in first-degree heart block, as compared to the number of T waves in first-degree heart block.

7. _____ The effect of increasing the preload on HR

8. _____ The effect of increasing the ESV on the preload

9. _____ The effect of Lanoxin on SV

10. _____ The effect of decreasing the concentration of the interstitial fluid proteins on the subsequent amount of fluid filtered out of a capillary

11. _____ The effect of Diovan on angiotensin II synthesis

12. _____ The likelihood of someone having septic shock, as compared to the likelihood of having anaphylactic shock

13. _____ The effect of stimulating beta receptors on the heart on the slope of the prepotential

14. _____ The nearness of the epicardium to the heart muscle, as compared to the nearness of the serous pericardium to the heart muscle

15. _____ Volume of the EDV, as compared to the volume of the ESV

16. _____ The effect of Inderal on the ESV

CHAPTER 8

Water Regulation and Kidney Function

I. **Intake:** The amount of water in the body does not change much because there is a balance between the water intake and water loss.
   a. **Sources**
      i. **Preformed:** The kind of water found within foodstuff.
         (1) Example: A peach has a lot of preformed water, but there is no preformed water in a rice cake.
      ii. **Metabolic:** All food has this kind of water; this is the kind of water you get when food is metabolized in the body. (Some animals can exist solely on preformed and metabolic water sources.)
      iii. **Fluids:** Examples—milk, water, Coke
         (1) **With Coke you get preformed water and metabolic water** because there is sugar in a Coke (as opposed to a Diet Coke).
   b. **Regulation:** Regulated by **sensation of thirst**
      i. **Osmoreceptors:** in the hypothalamus
         (1) Osmoreceptors are sensitive to the **saltiness of the blood** or when it is hypertonic (too salty); causes **the sensation of thirst**.
      ii. **Volume receptors:** monitor the extracellular fluid volume
         (1) When people lose a lot of blood (trauma); after an hour or so they get thirsty.
         (2) Volume receptors are located on the low pressure side (left atrium, pulmonary vein, superior or inferior vena cava, etc.).
         (3) ~**7% to 10% depletion of blood needed** before these are sensitive to the change in volume
      iii. **Renin-Angiotensin System**
         (1) Angiotensin II **vasoconstricts the arterioles** and **causes thirst** (along with other things).
      iv. Other: dryness of mouth; oral metering

II. Water loss

a. **Sources**
   i. **Unavoidable:** We do not regulate these to reduce water loss.
      (1) **Sweating:** The rate of sweating has nothing to do with maintaining water loss, but only with thermal regulation.
      (2) **Insensible evaporation:** Skin is porous—there is usually evaporation going on. In winter you can see your breath because you have humidified the air when you breathed it in. Thus, expired gases are humidified and represent some water loss.
      (3) **Feces:** Generally not too much water lost here
      (4) **Lactation**
      (5) **Tears**
   ii. **Urine:** The amount **varies according to the current state of hydration.**
      (1) If you are hypohydrated, then your body produces less urine (which is more concentrated).

b. **Regulation**
   i. **Antidiuretic hormone, or ADH (vasopressin):** When osmoreceptors in the hypothalamus are stimulated, they cause us to become thirsty and cause **the posterior pituitary** gland to release ADH, which acts at the kidney level (collecting ducts) to decrease the urine output.
   ii. **Aldosterone:** produced by the **adrenal cortex** (outside the adrenal medulla)
      (1) It produces several hormones, such as aldosterone.
      (2) It takes Na^+ out of the potential urine and puts it back in the blood.
      (3) Water follows Na^+, so aldosterone decreases urine production.
      (4) Aldosterone sends 3 Na^+ back into blood and 2 K^+ into urine.
      (5) **Things that stimulate aldosterone secretion:** low blood Na^+, angiotensin II, and elevated blood K^+

KIDNEY

I. Major functions

a. **Functions**
   i. Regulates the **ions** in the extracellular fluid (K^+, Ca^{++}, Na^+, etc.)
   ii. Has an important role in the **acid-base balance** (controls the pH of the urine and blood)
   iii. Regulates the **BP** via the renin-angiotensin mechanism.
   iv. **Waste removal:** urea, H^+ ions, etc.
   v. **Toxic compounds:** Waste can turn into toxic compounds; the kidney removes these.
   vi. Hormone production: **renin (from the JG cells of the afferent arteriole).**
   vii. **Degrades hormones**
   viii. **Osmotic pressure and volume:** regulates these in the extracellular fluid

b. **Anatomy**
   i. The two kidneys are bean-shaped, about 4" long and 1" thick.
   ii. They are located in the lumbar region of the back and **retroperitoneal** (behind the peritoneum and right against the abdominal wall, as if they are shrink-wrapped).
   iii. The **right kidney is more caudal.**
   iv. The adrenal glands are located above each kidney.

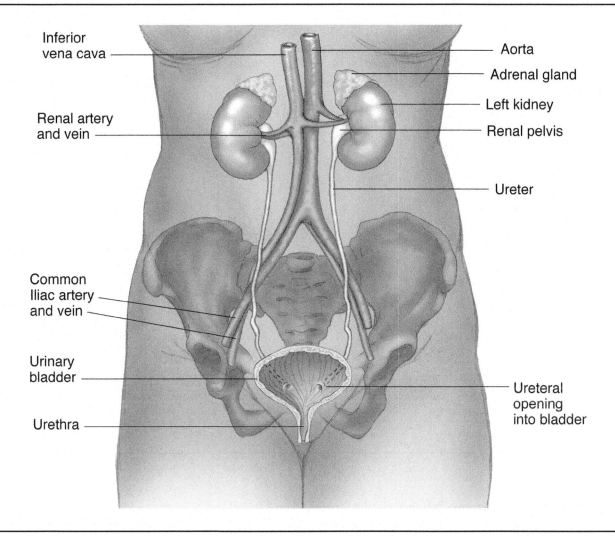

Inferior vena cava

Aorta

Adrenal gland

Left kidney

Renal artery and vein

Renal pelvis

Ureter

Common Iliac artery and vein

Urinary bladder

Urethra

Ureteral opening into bladder

FIGURE 8.1

II. Macroscopic anatomy
   a. If you cut a longitudinal section:
      i. The outer portion/side is the **cortex.** The cortex is very **well vascularized.** The interstitial fluid of the cortex is **always isotonic** with the plasma (300 milliosmoles).
     ii. Inside the cortex is the **medulla.** The medulla is **not as well vascularized.** The interstitium (interstitial fluid) is **progressively hypertonic** (more and more salty, about 1200 milliosmoles) as you move toward the center of the kidney.
   iii. The inner medulla contains the renal pyramids (triangular structures) and the renal pelvis, a hollow portion that leads to the ureter, then the bladder, and the urethra.

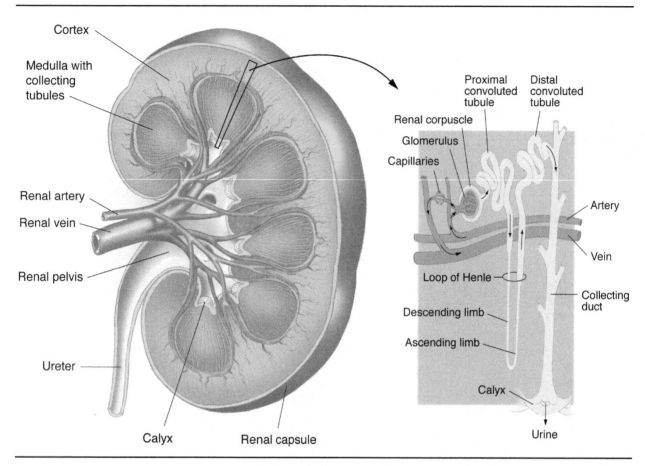

FIGURE 8.1

III. **Microscopic anatomy: Nephron**—the structural and functional unit of the kidney; there are ~1 million nephrons in each kidney. There are **two types of nephrons**: **juxtamedullary** (with long loops of Henle and vasa recta) and **cortical** (with short loops of Henle and no vasa recta).
   a. **Vascular Structures:** The kidney's vascular structures receive a significant portion of the cardiac output from the heart; about 23%–25% of all the cardiac output (at rest); the kidney helps to purify and clean the blood.
      i. **Renal Artery → Segmental Arteries → Interlobar Arteries → Arcuate Arteries → Interlobular Arteries** (more than one nephron can be served by these structures).
      ii. **Afferent Arteriole → Glomerular Capillary Bed → Efferent Arteriole → Peritubular Capillary Network** (around the tubular portion of the nephron) **→ Venous Drainage** (paired with veins).
      iii. This is a **Portal System**: There are two capillary beds in the series (one after another).
         (1) First is the **glomerular capillary bed**; second is the **peritubular capillary network**.
   b. **Tubular Structures**
      i. **Bowman's Capsule → Proximal Convoluted Tubule → Loop of Henle** (divided into different parts) **→ Descending Loop → Ascending Loop** (thick and thin portions) **→ Distal Convoluted Tubule → Collecting Duct** (serves more than one nephron) **→ Renal Pelvis → Ureter → Bladder → Urethra → Toilet**

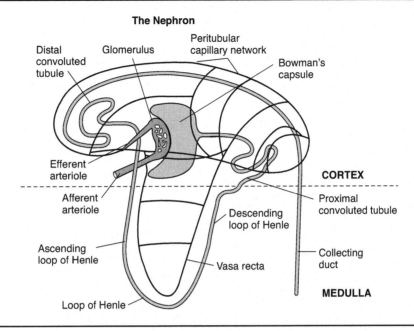

The Nephron

FIGURE 8.3

IV. Steps in urine formation
  a. **Three steps:**
  i. **Filtration:** Between the glomerulus and Bowman's capsule, filtration occurs as the **plasma moves from the glomerulus into Bowman's capsule**.
     (1) From the vascular portion to the tubular portion
     (2) Filter is somewhat indiscriminate (discriminates only by size) between good and bad stuff; both get filtered; thus, we need a mechanism to reabsorb the good stuff.
  ii. **Reabsorption:** from the tubular portion back to the peritubular capillary network (blood)
     (1) Movement of substances from the tubular portion to the vascular portion
  iii. **Secretion:** Some things may not get filtered, yet secreted from blood to tubular network.
     (1) Moving from the peritubular capillary network to the tubular portion of the nephron
  iv. These three steps are occurring all the time and are necessary for normal formation of urine.

V. Renal clearance: You cannot evaluate kidney function unless you look at the amount that is excreted and compare it to the plasma concentration of a substance.
  a. **Renal clearance: Hypothetical plasma volume containing the amount of any substance excreted in the urine per minute (notice the difference between excreted and secreted—do not confuse these)**
  b. Example: Feed two dogs the same amount of urea nitrogen and then collect the urine over 24 hours:

	Dog A	**Dog B**
Urea Nitrogen eaten:	14.4 g	14.4 g
Collected (24 hr):	14.5 ($\pm$.2) g	14.3 ($\pm$.2) g
Plasma concentration:	0.1 mg/ml	2.0 mg/ml

c. You cannot just look at the amount excreted; **you have to look at the plasma concentration.**
   i.   Dog B is sick!
d. Now say that both the dogs excrete 14.4 g per 24 hours, so that is 10 mg/minute.
e. **How many ml of Dog A's plasma does it take to contain 10 mg?**

   i.   Dog A: each ml = 0.1 mg
   ii. 10 ml = 1 mg

f. **What is the renal clearance of urea nitrogen of Dog A?**

g. **How many ml of Dog B's plasma does it take to contain 10 mg?**

h. **What is the renal clearance of Dog B?**

i. **Rules for calculating renal clearance:**
   i.   **Calculate the amount excreted and get units in mg/minute.**
   ii. **Calculate the plasma concentration and get units in mg/ml.**
   iii. **Divide #1 by #2.**
j. **Dog A:**
   i.   How much was excreted per minute?
   ii. Plasma concentration:
   iii. Divide #1 by #2:
k. **Dog B:**
   i.   How much was excreted per minute?
   ii. Plasma concentration:
   iii. Divide #1 by #2

VI. **Filtration:** Net movement of fluid through some semipermeable membrane according to the pressure gradient.
   a. **The pressures**
      Positive number: Favors filtration
      Negative number: Opposes filtration
      i.   **Blood pressure:** Pressure in the glomerulus is fairly high in respect to the other capillaries: **+60 mm Hg.**
           (1) Afferent arteriole has a greater diameter than the efferent arteriole.
      ii. **Osmotic pressure:** Plasma proteins are too big to be filtered and hence exert osmotic pressure (Na^+ doesn't because it is so small and thus filtered; RBCs cannot because they are too big): **-28 mm Hg.**
      iii. **Bowman's capsule pressure:** It has fluid in it, so there is hydrostatic pressure there: **-15 mm Hg**
           (1) Example: Prostate gland grows in older men and strangles the urethra. This would increase Bowman's capsule pressure and impair filtration.
      iv. Net filtration (normal) pressure is thus **+17 mm Hg (+60 – 28 – 15 = 17).**
   b. **The filter: glomerulus and Bowman's capsule**

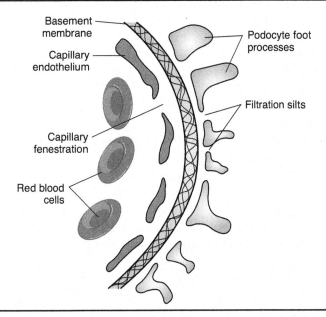

FIGURE 8.4

i. **Glomerulus:** A capillary with a simple layer of endothelium
    (1) **Fenestra:** Spaces between the adjacent endothelial cells
        a. Barriers to anything cell-sized or larger (RBCs are not filtered)
    (2) **Basement membrane:** Made up of **collagen and glycoproteins**; a net negative charge tends to repel a number of molecules such as albumin (plasma protein).
    (3) **Podocytes:** Epithelial cell layer of Bowman's capsule with finger-like projections; these projections partially overlap each other.
        a. Must go through the filtration slits in order to get through; if something is protein-sized or bigger, then it can't get from the glomerulus to Bowman's capsule.
    (4) Diseases can damage the membrane and allow other things to pass through.

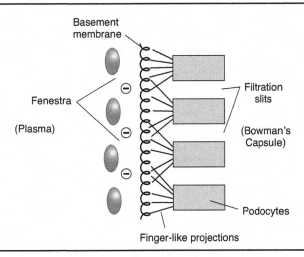

FIGURE 8.5

c. **Quantification of glomerular filtration rate:** Cannot measure according to the urine output, because some stuff gets reabsorbed.
  i. **Inulin:** Can measure GFR with this; it is freely filtered and inert to the rest of the tubule. It is neither reabsorbed nor secreted.
    (1) Infuse inulin so that plasma concentration is 1 mg/ml and the inulin collected in the urine is 125 mg/minute.
    (2) Where did the 125 mg/minute come from in the nephron? By filtration every minute; 125 ml of volume contains 125 mg.
    (3) **Renal clearance is 125; inulin's renal clearance is the same as the GFR, because the only way to get there is by filtration. The inulin clearance will always = GFR.**
    (4) Inulin is the gold standard to measure GFR. It is not naturally occurring though, so clinically we use creatinine.
  ii. Body makes something that is freely filtered and almost inert: creatinine.
    (1) **Creatinine clearance measures GFR clinically.**
  iii. **Compare the amount of inulin in the afferent arteriole versus the efferent arteriole:**
    (1) More in the _____ arteriole than the _____ arteriole.
    (2) Efferent arteriole: after it is filtered
    (3) Afferent arteriole: before it is filtered
    (4) **What is the concentration in the afferent and efferent arterioles?**

  iv. **Is *all* the fluid filtered if it's "freely filtered"? (That is, would there be inulin in the efferent arteriole? Why or why not?)**

  v. **Filtration fraction:** Percent of total renal plasma volume that enters Bowman's capsule.
  vi. **Hematocrit (HCT):** The portion of volume of whole blood that is due to the formed elements (RBCs, white blood cells, etc.)

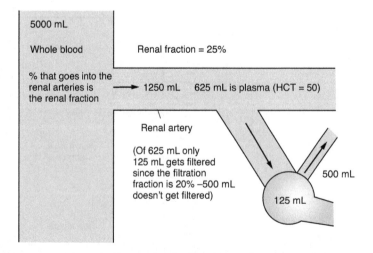

5000 mL

Whole blood   Renal fraction = 25%

% that goes into the
renal arteries is   → 1250 mL   625 mL is plasma (HCT = 50)
the renal fraction

Renal artery

(Of 625 mL only
125 mL gets filtered
since the filtration
fraction is 20% –500 mL   500 mL
doesn't get filtered)

125 mL

FIGURE 8.6

vii. Compare the amount and concentration of fibrinogen (plasma protein) in afferent and efferent arterioles.

Is this different from things that are filtered? Explain.

d. **Regulation of GFR**:

$$\text{GFR} = K_f \times (\text{net filtration pressure})$$

i. **Regulation of filtration pressure**: Regulating the **glomerular BP**

(1) **External mechanisms**: Mechanisms outside the kidney that regulate glomerular filtration rate

a. **Atrial natriuretic peptide (ANP or ANF)**: Cells within the atria of the heart produce this peptide hormone. ANP is released in response to the chronic stretching of the atrial walls.

- **ANP causes vasodilation of the afferent arteriole,** which will increase glomerular BP. The afferent arteriole increases the blood coming into the glomerulus, which increases the urine output by increasing GFR.

- **ANP inhibits renin secretion** (which affects angiotensin II) and **inhibits aldosterone secretion** (aldosterone takes Na^+ out of the filtrate and puts it back in the blood; water follows Na^+), which increases urine production.

- **ANP inhibits Cl^- and Na^+ reabsorption at proximal parts of the tubule.** This also increases urine production.

- **ANP inhibits ADH secretion,** which also increases urine output.

(**ANP also relaxes the mesangials, which does not change the pressure, but changes the filtration coefficient.**)

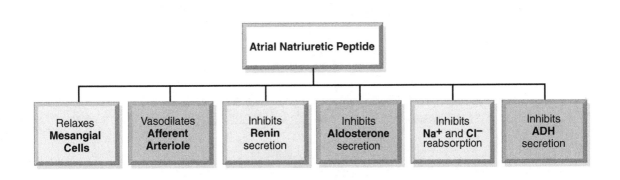

FIGURE 8.7

b. **Neuronal control** (nervous system control): The kidneys are richly innervated by the sympathetic nervous system.

—The sympathetic nervous system can **vasoconstrict the afferent arterioles**; this causes the glomerular BP to decrease and the GFR to decrease.

—Under sympathetic duress, the blood is shunted elsewhere besides the kidney.

(2) **Autoregulation**: regulation within the kidney itself

a. **Myogenic hypothesis**: The afferent arterioles have relatively elastic and muscular walls. If you increase the arteriole BP, then the walls contract back to self-regulate.

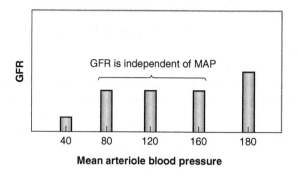

FIGURE 8.8

b. **Tubuloglomerular Feedback Hypothesis:** Focus on cells at the "crotch" of afferent arteriole, efferent arteriole, and distal tubule. Cells on the tubule here are the **macula densa**.

   —The macula densa and the efferent and afferent arterioles are known as the **juxtaglomerular apparatus (JGA).** The JGA monitors the GFR.

   —Anatomy of JGA:

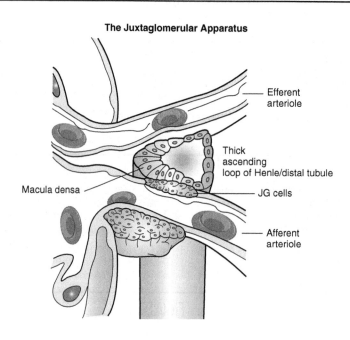

The Juxtaglomerular Apparatus

FIGURE 8.9

—Macula densa monitors the GFR by monitoring the amount of Cl^- in the filtrate.

—If the GFR is too high, then the rate of flow would be increased. Thus, there is less time to get Cl^- out of the urine, so by the time it reaches the macula densa, the

Cl⁻ concentration is too high. **The macula densa recognizes that the GFR is too high in this case.**

—If GFR is too low, then there would have been more time to remove a lot of Cl⁻ from the filtrate. **The macula densa recognizes that the GFR is too low in this case.**

—The macula densa cells have the capacity to communicate with the efferent and afferent arterioles. Mechanism is unknown.

—If the GFR is too low, the macula densa initiates a chemical messenger, which communicates with the afferent arteriole and makes it vasodilate. The glomerular BP increases and GFR is normal again.

—**GFR is kept constant despite changing arterial BP.**

(3) **Renin-angiotensin system**: stimulated by a decrease in NaCl in the urine (low GFR) or a decrease in blood volume

    a. Works opposite of **atrial natriuretic peptide.**

    b. Causes renin to be released from the granular cells (juxtaglomerular cells, or JG cells). Renin converts angiotensinogen into angiotensin I. This is converted by ACE into angiotensin II.

    c. Angiotensin II causes vasoconstriction everywhere (and especially at the efferent arterioles). **This increases GFR.** But how does this help the low blood volume?? Angiotensin II also…

    d. **Increases aldosterone secretion**

    e. **Causes thirst sensation**

    f. **Increases ADH secretion**

    g. **Increases Na⁺ reabsorption, independent of aldosterone, at the proximal tubule**

    Thus, overall angiotensin II increases GFR **and** increases BP.

ii. **Regulation of K_f**

(1) **Mesangial cells:** regulate blood flow into the capillaries of the glomerular capillary bed. When they contract they shut off some of the surface area of the capillary supply in the glomerulus. When they relax they increase the surface area of the capillary supply in the glomerulus.

    a. **Atrial natriuretic peptide (ANP) has its effect on the mesangial cells; it relaxes them to expose more surface area.**

(2) **Podocytes:** contractile properties; slits between the fingers of the podocytes can be altered to make them less or more permeable. The fingerlike projections are regulated to affect filtration at the podocyte level, but the mechanism is unknown.

VII. **Reabsorption:** GFR = 125 ml/minute; 7.5 liters an hour; 180 liters a day! Anything from the vascular portion to the tubular portion is destined to be excreted. We don't urinate 180 liters/day.
—99% of filtrate is also reabsorbed.

    a. **Organic reabsorption**

      i. **Proteins:** These are not supposed to get filtered in the first place because they are so large, but some albumin leaks through. There are mechanisms by which proteins are broken down to be reabsorbed.

      (1) **Proteinuria:** protein in the urine; happens if there is a problem in the kidney.

      ii. **Transport/tubular maximum mechanism:** A carrier protein transports a particular molecule (e.g., glucose). If there are too many molecules, then they saturate the carrier protein: there is a transport mechanism limitation. The rest must then be excreted. Normally, T_{load} is well below T_{max} for glucose (diabetes mellitus patients are the exception).

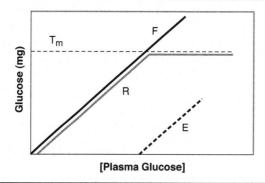

FIGURE 8.10

Note: Really lines F and R would be directly on top of each other until exceeding T_{max}.

- a. **Tubular/transport maximum mechanism limit is independent of the plasma concentration.**
    - i. R = amount of glucose that is reabsorbed.
    - ii. F = amount of glucose that is filtered.
    - iii. E = amount of glucose that is excreted.
    - iv. $Tm = T_{max}$, or maximum amount of glucose that can be reabsorbed per minute.
- b. **Tubular load:** total amount (mg) of a substance that enters the tubular portion of the nephron each minute (amount of substance from the glomerulus to Bowman's capsule)
    - i. Two variables: GFR & plasma concentration
    - ii. $T_{load} = GFR \times Plasma\ concentration$
    - iii. $T_{max} = GFR \times Renal\ plasma\ threshold$.
- c. Glucose is filtered, even past Tm. Glucose is reabsorbed to this point (Tm), but then it gets excreted.
    - i. As long as the tubular load is less than Tm, the glucose gets reabsorbed. Once you exceed Tm, the excess doesn't get reabsorbed; it goes into the urine.
- d. **Diabetes mellitus:** There is a problem with the insulin in the body; some people don't produce insulin (type 1) and some don't respond to insulin (type 2); the insulin does not affect the body like it is supposed to.
    - i. **In normal people:** Insulin promotes the transport of glucose from the extracellular fluid to the intracellular fluid.
    - ii. **In diabetics:** The **extracellular fluid remains laden with glucose**, because the insulin-mediated system of taking up glucose into cells is not working correctly.
    - iii. **The tubular load exceeds the Tm and excess glucose is excreted in the urine.** Normally it is all reabsorbed in the proximal convoluted tubules—but not in diabetics.
        —**Glucosuria:** glucose in the urine (also called **glycosuria**)
- 3. **Renal plasma threshold:** That plasma concentration of a substance that results in a tubular load equal to the T_{max} (assumes constant GFR).

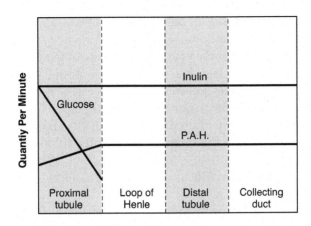

FIGURE 8.11

b. **Reabsorption of selected molecules**
   i. **Inulin:** freely filtered—gets in and stays in
   ii. **Insulin:** a large protein; doesn't get in the tubules at all. This would be a straight line across the bottom of the above graph.
   iii. **Glucose:** gets in the tubules. Normally the tubular load is less than the T_{max}, and it is all reabsorbed at the proximal convoluted tubule.
   iv. **Creatinine:** Behaves like inulin.
   v. Para-aminohippurate (**PAH**): Is being **secreted** (above we see more is coming into the tubules at the site of the proximal tubule).
   vi. **Glucose and amino acids are reabsorbed in the tubules by the T_{max} limited mechanism.**

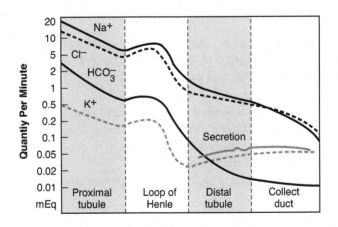

FIGURE 8.12

c. **Ionic Reabsorption**
   i. **Sodium:** Most of the Na^+ is actively reabsorbed in the proximal convoluted tubule; it passively diffuses back in at the descending loop of Henle. It is then actively transported

out at the ascending loop. Around 8% is left at the distal tubules and collecting ducts. **This last 8% is dependent on aldosterone.** The other 90%–92% is not regulated.

   ii. **Chloride:** This basically follows Na^+ around.

   iii. **Potassium:** is reabsorbed at the proximal convoluted tubule (actively). It diffuses in passively at the descending loop of Henle and is actively transported out at the ascending loop. The secretion at the distal tubules and collecting ducts can be regulated by **aldosterone.**

   iv. **Calcium:** The kidney is the major regulatory organ for Ca^{++} (filter discriminates on the basis of size). Ca^{++} is freely filtered but less than 2% appears in the urine.

      (1) Reabsorption occurs in the proximal convoluted tubules for about 60% of Ca^{++} (is not regulated).

      (2) Remaining reabsorption (40%) occurs in the distal tubules and is regulated by **parathyroid hormone.**

   v. **Phosphate:** More like glucose; at low concentrations it is all reabsorbed by the T_{max} limited mechanism.

      (1) If the plasma concentration is significantly high, then the excess is excreted. If the plasma concentration is significantly low, then phosphate ions are reabsorbed.

   vi. **Aldosterone Regulation:** Regulates the Na^+ in the distal tubule and the collecting duct (8-10% of the Na^+ is still around at this point).

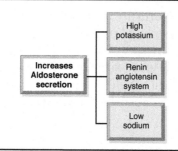

FIGURE 8.13

      (1) Na^+ reabsorption and K^+ secretion: 3 Na^+ reabsorbed for 2 K^+ secreted

      (2) Aldosterone is produced by the adrenal cortex in response to **high plasma K^+ levels, low Na^+ levels, and angiotensin II.**

VIII. **Osmotic stratification:** The mechanism by which urine is concentrated. The progressively hypertonic medulla affects the concentration of the urine. This occurs due to a phenomenal reaction between the descending limb and the ascending limb of the loop of Henle.

  a. **Facts about ascending and descending limb:**

   i. **Descending limb** of the loop of Henle is freely permeable to ions and water; there is no active transport.

   ii. **Ascending limb** of the loop of Henle has active transport of ions against the concentration gradient. The greater the gradient, the more difficult it is to transport ions. The ascending limb is also impermeable to water.

      (1) The greatest difference that can exist between the ascending limb fluid compartment and the extracellular fluid compartment is 200 milliosmoles.

  b. **Mechanism**

Schematic illustrating how urine is diluted in the descending and ascending loops of Henle. In this cartoon we have attempted to illustrate the principle of how filtrate dilution happens in the nephron. Note that we will progress from A to F by either "turning on the pumps" that move ions in or out of tubules OR "moving the fluid" through the fluid on its way to being excreted from the body, as designated by arrows, similar to the path it would take in a kidney. Note that we have drawn the columns next to each other for simplicity (without including the extracellular space that is really there). Also note that we have not paid any attention to specific ions or pumps. This is intended to teach the major principle of nephron dilution of filtrate. First, consider fluid that enters the descending loops is about 300 milliosmolar (mOsm), similar to plasma (A). Next (B), we turn "on" the pumps that move ions out of the tubule. Because the descending column is permeable to water, it will simply equilibrate with what is in the extracellular space around it if that is more concentrated. The ascending loop, however, pumps out ions, and retains water. Thus, we see that when we turn on the pumps, the ascending column pumps ions out to become more dilute while the descending pump simply allows those "pumped out" ions to equilibrate with inside the tubule, thus requisitely becomes more concentrated (in reality we can consider fluid moving out of the descending tubule for this equilibration with its surrounding extracellular space). We can then move the fluid (C) and again turn on the pumps (D). And again in (E) and (F). The end result is 2 fold; filtrate enters the collecting ducts as a very dilute urine, and the kidney has become "progressively hypertonic" as you move toward the middle of it. We now need to only "move filtrate" through the middle of the kidney and insert and remove water channels to regulate fluid balance (osmosis does the work). As you will see, ADH regulates the number of water channels in the collecting duct for exactly this purpose. Note that the human kidney can only concentrate up to a certain difference, thus limiting our ability to ingest fluids above a certain concentration. For example, consider that sea water is too concentrated to drink.

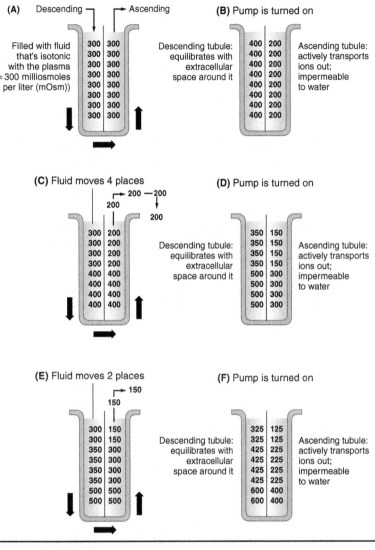

FIGURE 8.14

(1) The concentration is greater at the bottom and is also the same as the interstitium of the medulla of the kidney (interstitial fluid = hypertonic).
(2) We're trying to concentrate the filtrate, but so far we have only diluted it!! The filtrate runs through the medulla and the interstitial fluid becomes progressively hypertonic.
(3) **We now have the osmotic potential to remove the water from the filtrate as it passes back through the medulla via the collecting ducts.**

IX. **Volume reabsorption:** The collecting duct permeability depends on the presence of ADH (decreases the urine volume). **ADH works on the epithelium membrane of the collecting ducts to reabsorb water.** If the ADH is low, then the urine that is excreted is not concentrated; if the ADH is high, then the urine that is excreted is concentrated. **ADH is produced in the hypothalamus and secreted into the blood by the posterior pituitary gland.**
   a. **Diabetes:** People who suffer from **diabetes have a diuresis.**

    i. **Diabetes insipidus:** People with this form of diabetes cannot make enough ADH. If there is no ADH, then they will excrete a lot of urine (**~15% to 20% of water reabsorption is dependant on ADH;** the rest is obligatorily reabsorbed).

        (1) Treat by giving patient ADH.

    ii. **Diabetes mellitus:** People who suffer from this form of diabetes have a diuresis, but for a different reason. They have elevated plasma glucose levels.

        (1) **The tubular load of glucose exceeds the T_{max}** so not all the glucose gets reabsorbed.

        (2) The difference between the tubular load and the T_{max} gets excreted.

        (3) **Glucosuria:** Glucose **osmotically pulls the water** into the tubule. It is excreted along with the glucose.

        You are stranded on an island with two friends. They both have a severe diuresis. One has diabetes insipidus and the other diabetes mellitus (but you do not know which is which). You have one insulin injection that will save the friend with diabetes mellitus. How do you know, with no equipment at your disposal, to whom to give the injection?

b. **Anatomy:** There is an anatomical difference between the ascending and descending loop of Henle.

    i. **Ascending:** has a thin portion at the bottom and a thick portion at the top

    ii. **Descending:** mostly thin

    iii. There is a larger concentration in the interstitium **at the bottom of the loop of Henle (due in part to the concentration of urea).**

c. **Urea:** This is freely filtered and moves back and forth between the tubule and the interstitium, except at the thick portion of the loop of Henle and in the first region of the collecting duct, where urea is not permeable. At this point, water still can move out (early part of collecting duct), so urea becomes more concentrated. But, as fluid moves down the collecting duct (from the cortical collecting duct to the medullary collecting duct, and water is being reabsorbed), the urea concentration in the tubule is even greater, eventually causing a second reabsorption into the interstitium.

d. **If a substance is reabsorbed, it will always have a renal clearance less than a simultaneous inulin measurement.**

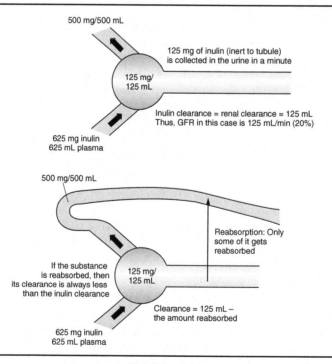

FIGURE 8.15

X. Secretion: If a substance is secreted, it will always have a renal clearance greater than a simultaneous inulin measurement.

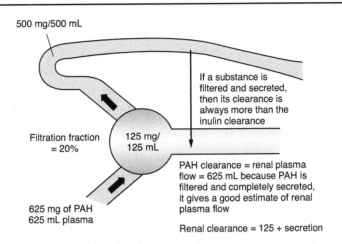

FIGURE 8.16

a. **Examples:** K^+, uric acid, bile salts, certain drugs (penicillin), H^+, certain food additives, PAH
   i. **K^+** is filtered, reabsorbed, and secreted.
   ii. **Uric acid** is filtered, reabsorbed, and secreted. It is made by the body (when purines are metabolized).
      (1) Uric acid is filtered, reabsorbed, and secreted.
      (2) Treat disorders by using a **uricosuric:** blocks reabsorption; you are left with what's filtered and secreted.
      (3) Renal clearance of **uric acid:** greater than or less than **inulin?**
b. Value of **(PAH):** completely cleared from the renal blood flow in a single pass through the kidney (because it is secreted so well).
   i. **Does something special: It measures the renal plasma volume.**
c. Drug **(PROBENECID):** blocks reabsorption of uric acid
   i. **Gout:** uric acid levels are too high. **Uric acid precipitates out into crystals** in the feet and hands. Very painful!
   ii. **(PROBENECID) is a uricosuric.** It blocks the reabsorption of uric acid and thus increases uric acid in the urine.
   iii. **(PROBENECID) can also be used for venereal disease** because it blocks the secretion of penicillin-like drugs. Penicillin is actively secreted; Probenecid keeps penicillin at high levels in the blood.

XI. Pathologies
   a. **Renal failure**
      i. Types
         (1) Chronic
         (2) Acute
      ii. Causes
         (1) Infections
         (2) Toxins

(3) Immune responses

(4) Obstructions

(5) Blood supply

    b. Proteinuria

    c. Nocturnal enuresis

XII. Diuretics

    a. Uses: Cardiac failure, congestive heart failure, liver failure (edema), hypertension. Use anytime we need to decrease the extracellular volume.

    b. Examples

      i. Nonpharmacological

        (1) Methylxanthine: Phosphodiesterase inhibitors (**caffeine**) thought to be diuretics (increase urine output). **The mechanism is poorly understood,** and new evidence argues that they actually are no diuretics *per se* (see work of Lawrence E. Armstrong for more on this topic). A proposed mechanism is that they increase the glomerular BP to increase GFR. Another proposed mechanism is that they inhibit the reabsorption of Na^+ in the tubules.

        (2) ETOH (ethanol): A CNS depressant. **This inhibits the release of ADH.**

      ii. Pharmacological

        (1) Mannitol (OSMOTROL): The substance is filtered and inert to the tubules. It is of such a size that it holds water in the tubules by exerting an osmotic pressure.

        (2) Thiazides ("low ceiling" diuretics): inhibit Na^+ reabsorption at the distal parts of the tubule.

          —Hydrochlorothiazide (HYDRO-DIURIL)

        (3) Loop diuretics ("high ceiling" diuretics): potent diuretics that work in the loop of Henle. These inhibit the pumping of ions out of the ascending tubule. Thus we get a "double whammy"—**the ions exert an osmotic pressure to hold more water in the tubule *and* the progressively hypertonic interstitium is not created to the same degree.**

          —Furosemide (LASIX)

        (4) Potassium sparing: Because diuretics tend to cause a loss of K^+, these work to spare K^+ (keep it in blood) by the fact that they are aldosterone antagonists. Remember that aldosterone reabsorbs 3 Na^+ and secretes 2 K^+. Thus, we have diuresis and maintain K^+.

          —(ALDACTONE)

        (5) Combinations between K^+ sparing and thiazide diuretics:

          —(DYAZIDE)

        (6) People on diuretics have to monitor K^+. Some K^+ supplements: (**K-DUR, SLOW-K, MICRO K, and KLOR-CON**)

SUGGESTED READINGS

1. Guyton, A.C., and Hall, J.E. 2011. Chapters 26–31 in *Textbook of Medical Physiology*, 12th ed. Baltimore: Saunders.

2. Koeppen, B.M. and Stanton, B.A. 2008. Chapters 32-35 in *Berne & Levy Physiology*, 6th ed. St. Louis: Mosby.

CHAPTER 9

Acid/Base Balance

I. **Introduction**
   a. **Acids & Bases:** Bronsted acids (H^+-containing substances that separate in solution to form H^+ ions and anions)
   b. Express this in concentration (of H^+) & pH values ($-\log [H^+]$).
   c. Effects of altered state: In the plasma the pH is 7.38–7.42. This range is remarkably stable. The fluid in the stomach has a pH of 1 to 2.
      i. **Acidosis:** Too many H^+ ions; people who suffer from this grow progressively lethargic until they become comatose and may die. As the pH falls, there is an **increase in ionized Ca^{++}** available. This free Ca^{++} blocks the voltage-gated Na^+ channels.
      ii. **Alkalosis:** As pH increases, there is a **decreased amount of extracellular Ca^{++}**. As it declines (less blocking of Na^+ channels), we get more action potentials. People suffering from this experience a tingling sensation in their extremities until they get tetanic contractions and eventually die.
   d. **Sources of Acid and Bases:** Carbonic acid is formed as a result of aerobic respiration (because CO_2 is formed). At rest we produce 12,500 milliosmoles per day. If this amount of carbonic acid were placed into a volume of water equal to the plasma volume, it would cause $[H^+]$ to be $\sim$100 million times normal.
      i. $CO_2 + H_2O \leftarrow$ (**carbonic anhydrase**) $\rightarrow H_2CO_3 \longleftrightarrow H^+ + HCO_3^-$
      ii. We are essentially **always fighting acidosis**.
      iii. We also **produce acids by breaking down nutrients** and **organic acids from metabolism** (lactate, etc).
      iv. Bases typically not a problem, although some foodstuffs produce basic groups.

II. Defenses
   ### a. Chemical Buffer Systems
   i. Definition: Consist of a pair of substances involved in a reversible reaction, one of which can yield free H^+ as the pH begins to rise and the other that can bind H^+ as the pH begins to fall.

   ii. Examples
   (1) Bicarbonate
   (2) Protein
   (3) Phosphate
   (4) Hemoglobin

   iii. Henderson-Hasselbach: pH = pKa + log (base/acid)
   (1) $pH = pKa + \log (HCO_3^-/CO_2)$
   (2) $pH = 6.1 + \log (HCO_3^-/CO_2)$
   (3) $pH = 6.1 + \log (20/1) =$ **7.4 (normal)**

   iv. Examples
   (1) **Bicarbonate:** most important extracellular buffer system
   $$CO_2 + H_2O \leftarrow CA \rightarrow H_2CO_3 \longleftrightarrow H^+ + HCO_3^-$$
   CA = carbonic anhydrase. The law of mass action dictates which way the reaction goes. If we increase H^+ ions, the reaction will go to the left. This is a good buffer system for acidosis. You cannot use this equation for respiratory acidosis (only metabolic acidosis) because respiratory acidosis cannot be fixed by breathing off CO_2 (metabolic acidosis *can* be compensated for this way).

   (2) **Proteins:** Most important intracellular buffers. Amino acids act like proton donors or proton receivers.

   (3) **Phosphate** (NaH_2PO_4: Na_2HPO_4): Phosphate is not in high concentrations in the body.
   $$HCl + Na_2HPO_4 \longleftrightarrow NaH_2PO_4 + NaCl$$
   Not powerful because it's not in high concentrations. **This system is important in the kidneys** (because of high concentrations) but not important in the blood.

   (4) **Hemoglobin:** H^+ ions can be buffered by hemoglobin. Example: Buildup of CO_2. This CO_2 comes into the RBC and is hydrated to form H_2CO_3 (which then goes to HCO_3^-). H^+ ions cannot just "hang out" in the RBCs, so they get buffered as shown below.

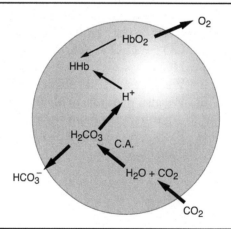

FIGURE 9.1

b. **Respiratory System:** This system acts in **minutes to hours** (not quite as fast as the chemical buffering systems).
    i.   If one becomes more acidotic, then they simply blow off more CO_2 to compensate. If one is alkalotic, then they blow off less CO_2 to lower the pH.
c. **Kidneys:** Kidneys regulate **long term**. They have the ability to alter the pH of the urine according to the status of the H^+ ions. There are many mechanisms by which we regulate pH long term via the kidneys.
    i.   The **bicarbonate ion is freely filtered** (with Na^+) and would be lost in the urine unless re-absorbed. We don't want to excrete HCO_3^- because we are usually fighting acidosis. We need the bicarbonate back in the blood to buffer. **There is no transport system in the tubule for transporting HCO_3^- directly. So one way…**
    ii. Secrete H^+ in exchange for Na^+. H^+ combines with the bicarbonate ion to form carbonic acid. Then in the presence of carbonic anhydrase this becomes CO_2 and H_2O. H_2O gets excreted in urine and CO_2 goes back to the cell to produce more HCO_3^-.
    iii. Indirectly using CO_2 to transfer the bicarbonate back into the blood since there is **no direct mechanism to reclaim HCO_3^-.**
    iv. In this case, urine is slightly acidic.

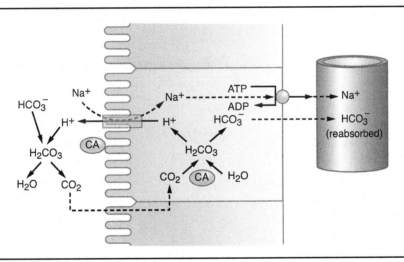

FIGURE 9.2

In Figure 9.2 we see the indirect transfer of filtered bicarbonate to bicarbonate in the blood. Notice the active transport mechanism to get bicarbonate into the blood. If we were to have the need to fight alkalosis, we would then begin excreting the filtered bicarbonate (more basic urine).

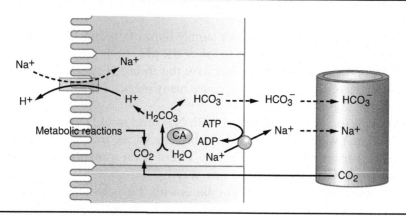

FIGURE 9.3

In Figure 9.3 we see bicarbonate being made *de novo* (new) from metabolic CO_2 and then being reabsorbed into the blood. Also note the secretion of H^+. We can secrete down to a pH of 4.5.

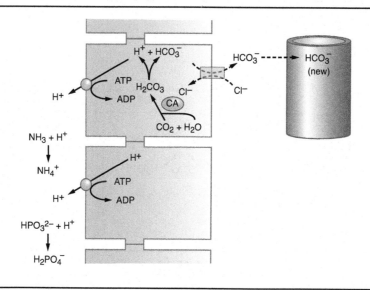

FIGURE 9.4

Figure 9.4 shows ways to buffer H^+ in the filtrate. As we secrete H^+ into the filtrate to a pH of 4.5, it then becomes necessary to "buffer" protons in the filtrate so we can continue to secrete them. Here the kidney uses both phosphate and ammonia to buffer the secreted H^+.

Buffering systems in the filtrate:

$NH_3 + H^+ \rightarrow NH_4^+$

$HPO_4^{2-} + H^+ \rightarrow H_2PO_4^-$

III. Acid-Base Imbalances

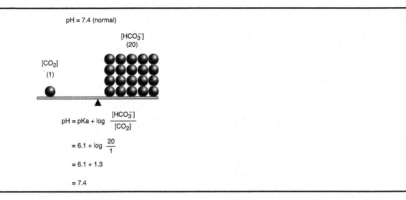

FIGURE 9.5

Figure 9.5 represents the normal pH balance in the blood. Notice the Henderson-Hasselbach equation; *the ratio* of HCO_3^- to CO_2 is what must be maintained. The exact concentrations could be changed (e.g., 10:0.5) without a concomitant change in pH.

a. **Respiratory acidosis:** The problem is in the respiratory system. This is caused by an **excessive retention of CO_2 (not breathing enough).**

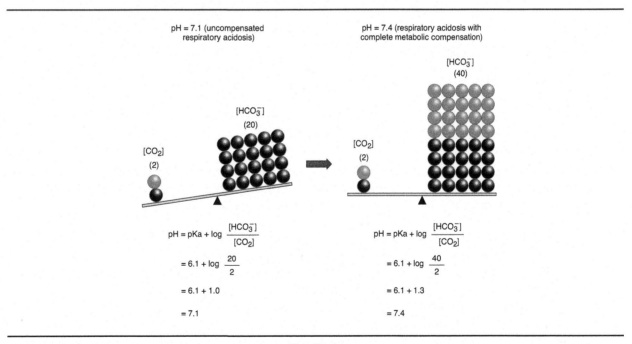

FIGURE 9.6

In Figure 9.6, we see that we can solve the problem at the kidney level by making bicarbonate *de novo* and reabsorbing all of the existing bicarbonate. We would also secrete protons (to a pH of 4.5 or up to what is needed) to be excreted.

i. **Causes:** Lung disease, depression of the respiratory centers in the brain, obstruction in the respiratory pathways (asthma, emphysema, pneumonia). **The bottom line with this is that somehow the ability of the patient to get rid of CO_2 is impaired.**

ii. As we accumulate CO_2 the pH drops.

iii. We cannot correct it by blowing off CO_2 because the respiratory system is impaired. **The kidney must be the solution.**

iv. **Solution:** Retain all the HCO_3^- that gets filtered and make HCO_3^- *de novo* until the pH returns to normal. The pH of the urine is lowered and HCO_3^- increases in the plasma; the ratio is restored to 20:1 **(the ratio is *always* what is important).**

b. **Respiratory alkalosis:** Not likely to see this very often. **Due to too much CO_2 being blown off (breathing too much).**

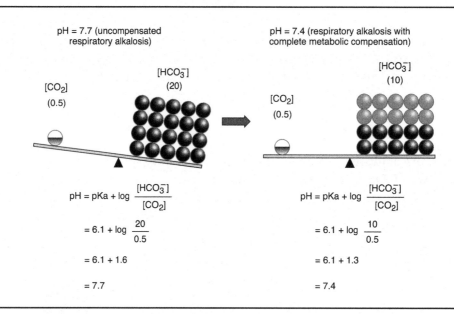

FIGURE 9.7

In Figure 9.7, we see that we can solve the problem at the kidney level by excreting bicarbonate. We would also retain protons in the blood (as the pH of the urine may get as high as 8.5 or up to what is needed).

i. **Causes:** Hyperventilation, anxiety, and high altitudes (too much CO_2 breathed off because chemoreceptors sense low PO_2 and increase breathing rate)

ii. The body corrects this by excreting HCO_3^- instead of reclaiming it. The urine pH increases (up to 8.5).

iii. **Solution: Nonphysiological**—Have patient breathe into a paper bag to rebreathe CO_2 and restore the pH. **Physiological**—The kidney will correct by excreting bicarbonate.

c. **Metabolic acidosis:** Likely to encounter with a patient. **Acidosis is for a reason other than the respiratory system** (e.g., lactate accumulation from strenuous exercise, diabetes mellitus [keto acids], too much aspirin, antifreeze ingestion, or severe diarrhea).

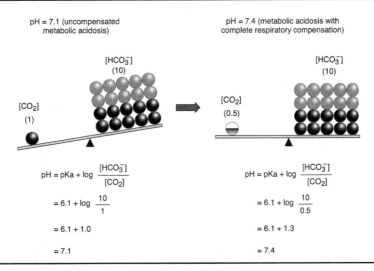

FIGURE 9.8

In Figure 9.8, we see that we can solve the problem at the kidney level by retaining bicarbonate (and making bicarbonate de novo) *and* by blowing off more CO_2 at the lung level. We would also secrete protons into the filtrate.

   d. <u>Metabolic alkalosis:</u> Not common; due to the ingestion of many alkalotic drugs (antacids) or with excessive vomiting. **Solution:** breathe less to retain more CO_2 and have the kidneys excrete bicarbonate.

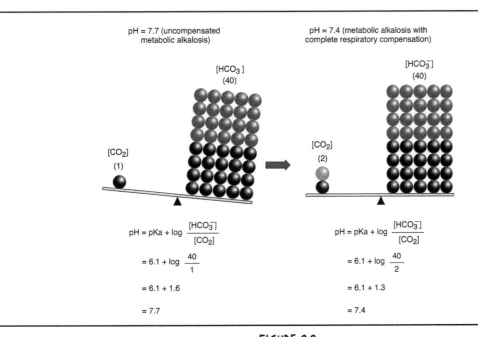

FIGURE 9.9

In Figure 9.9, we see that we can solve the problem at the kidney level by excreting bicarbonate *and* at the lung level by breathing less to retain CO_2. We would also retain protons in the blood (as the pH of the urine may get as high as 8.5 or up to what is needed).

IV. A final tutorial on Acid-Base and the Anion Gap (AG)

Clinically we will use:

Anion Gap (AG) = [Na$^+$] - ([Cl$^-$] + [HCO$_3^-$])

Usually this is 10–12 mM.

If it is *higher* than 10–12, it can mean that there is some unmeasured anion in the blood (e.g. lactate).

Check serum albumin (SA), as a fall in SA of just 1g/dL can mean anion gap decreases by 2.5 mEq/L.

Usually a high AG is due to some anion in higher than normal concentrations (lactate, acetoacetate). With this, you see low HCO$_3^-$.

An easy way to approach acid-base problems:

Consider the following values "normal" for arterial blood:

pH = 7.38-7.42

pCO$_2$ = 35-45 Torr

[HCO$_3^-$] = 22-26 mM

Because alkalosis and acidosis refer to the "process" we can consider respiratory and metabolic acidosis/alkalosis separately. The pH tells us the resultant status of the blood; it can be normal or there can be acidemia or alkalemia.

Steps to acid-base problems:

a. Remember that there is a metabolic and respiratory component to each acid-base problem. They can both be causing the problem (e.g. both causing acidosis) OR one can be compensating for the problem that the other is causing (e.g. respiratory acidosis with a "metabolic alkalosis" in an attempt to compensate). However, when the body tries to compensate for the problem, it NEVER overcompensates!

b. Look at the pH first; this tells you if it is acidic or basic.

c. Look at the pCO$_2$ and determine if it is LOW or HIGH. This will tell you if the breathing is the problem or an attempt to compensate. For example, if you see a pH of 7.46 and a pCO$_2$ of 50 you know that you have alkalosis but a respiratory acidosis in an attempt to compensate!

d. Look at bicarbonate concentration ([HCO$_3^-$]) to determine the metabolic status of the patient. Usually, when acidemia is present, you will see low bicarbonate, as it is being "used up" to buffer the excess protons.

e. Compensation can be "complete" or "incomplete" depending on whether it puts the pH back into the "normal" range.

For each arterial blood gas (ABG), think of your "10 abnormal choices":

a. Metabolic and respiratory acidosis.

b. Metabolic acidosis with respiratory (alkalosis) compensation.

c. Respiratory acidosis with metabolic (alkalosis) compensation.

d. Metabolic acidosis.

e. Respiratory acidosis.

f. Metabolic and respiratory alkalosis.

g. Metabolic alkalosis with respiratory (acidosis) compensation.

h. Respiratory alkalosis with metabolic (acidosis) compensation.

i. Metabolic alkalosis.

j. Respiratory alkalosis.

EXAMPLES:
pH = 7.33
pCO_2 = 50 Torr
$[HCO_3^-]$ = 20 mM
ANSWER:
This is both respiratory and metabolic acidosis.

pH = 7.34
pCO_2 = 50 Torr
$[HCO_3^-]$ = 27 mM
ANSWER:
This is respiratory acidosis with incomplete compensation by metabolic alkalosis. How do you know it is not a metabolic alkalosis with respiratory acidosis to compensate? Recall rule 1 above – the body never overcompensates! You know the compensation component is the component that does not agree with the pH! Here the pCO2 is high and pH low, so that that is the primary problem. By itself, high bicarbonate causes a high pH, so we know this must be a compensatory mechanism for the low pH.

pH = 7.45
pCO_2 = 23 Torr
$[HCO_3^-]$ = 22 mM
ANSWER:
This is respiratory alkalosis.

pH = 7.46
pCO_2 = 44 Torr
$[HCO_3^-]$ = 31 mM
ANSWER:
This is metabolic alkalosis.

pH = 7.33
pCO_2 = 32 Torr
$[HCO_3^-]$ = 17 mM
ANSWER:
This is metabolic acidosis.

pH = 7.39
pCO_2 = 33 Torr
$[HCO_3^-]$ = 19 mM
ANSWER:
This is metabolic acidosis with complete compensation by respiratory alkalosis.

pH = 7.41
pCO_2 = 46 Torr
$[HCO_3^-]$ = 29 mM
ANSWER:
This is metabolic alkalosis with complete compensation by respiratory acidosis.

pH = 7.39
pCO_2 = 41 Torr
$[HCO_3^-]$ = 24 mM
ANSWER:
This is normal.

Rather than speaking of "complete" and "incomplete compensation", a more advanced way is to look at what the EXPECTED changes in bicarb and pCO_2 should be in a "functional" system. In other words, if you look at the bicarb changes in metabolic acidosis and the respiratory changes are LESS than you would expect, you might conclude that there is metabolic acidosis AND a respiratory acidosis (based on the response being LESS than it should be based on below criteria).

To make estimations of "appropriate" responses, consider the following:
Metabolic Acidosis: For each decline of 4 in bicarb, a decline of 5 in pCO_2 is appropriate.
Metabolic Alkalosis: For each increase of 10 in bicarb, an increase of 6 in pCO_2 is appropriate.
Respiratory Acidosis: For each increase of 10 in pCO_2, an increase of 1 in bicarb is appropriate (4 if chronic).
Respiratory Alkalosis: For each decrease of 10 in pCO_2, a decrease of 2 in bicarb is appropriate (4 if chronic).

Finally, the following estimation can be quite helpful clinically when there are "combined disorders":
Delta-delta = (e.g. change in AG) / (e.g. change in bicarb)
If this is 1.0, then it is a pure "AG acidosis" (e.g. lactate build up)
If this is less than 1.0, then there is both "AG acidosis" and "non-AG acidosis" (e.g. lactate + diarrhea)
If this is greater than 1.0, then there is "AG acidosis" and "metabolic alkalosis" (e.g. diabetic ketoacidosis + vomiting)

e. **Quick acid/base review:**
$H^+ + HCO_3^- \longleftrightarrow H_2CO_3 \longleftarrow(CA)\rightarrow CO_2 + H_2O$
1) Metabolic acidosis = you attempt to correct with breathing.
2) Respiratory acidosis = you cannot correct with breathing (imbalance is *because* of breathing).

(The same goes for alkalosis.)

Breath rapid → breath off CO_2 → H^+ decreases, pH increases (respiratory alkalosis)
Breath slow → retain CO_2 → H^+ increases, pH decreases (respiratory acidosis)
Add strong acid → H^+ increases, pH decreases (metabolic acidosis)
Add strong base → H^+ goes down, pH increases (metabolic alkalosis)

SUGGESTED READINGS

1. Guyton, A.C., and Hall, J.E. 2011. Chapter 30 in *Textbook of Medical Physiology*, 12th ed. Baltimore: Saunders.
2. Koeppen, B.M. and Stanton, B.A. 2008. Chapter 36 in *Berne & Levy Physiology*, 6th ed. St. Louis: Mosby.

CHAPTER 8 AND 9 QUESTIONS

1. _____ Choose the sequence of structures that a single molecule of inulin could encounter.
   a. Afferent arteriole, peritubular capillary network, renal vein
   b. Interlobular artery, proximal convoluted tubule, ascending loop of Henle, renal pelvis
   c. Glomerulus, efferent arteriole, distal convoluted tubule, ureter
   d. All of these
   e. Two of these

2. _____ Angiotensin II
   a. Causes the secretion of ANP
   b. Enhances the secretion of ADH
   c. Directly increases the resorption of sodium
   d. All of these
   e. Two of these

3. _____ Choose the sequence of structures that a single molecule of FSH could encounter. Only the sequence is important.
   a. Segmental artery, glomerulus, descending loop of Henle, collecting duct
   b. Interlobar artery, arcuate artery, efferent arteriole, vasa recta
   c. Glomerulus, afferent arteriole, peritubular capillary network, renal vein
   d. Two of these
   e. None of these

4. _____ As person suffering from respiratory alkalosis attempts to compensate, he or she
   a. Reduces the bicarbonate that is filtered
   b. Excretes a basic urine
   c. Likely excretes disodium hydrogen phosphate
   d. All of these
   e. Two of these

5. _____ People with respiratory acidosis
   a. Typically breathe faster as they compensate
   b. May also be being treated with SEREVENT
   c. May secrete H^+ from the peritubular capillary network into the proximal convoluted tubule.
   d. All of these
   e. Two of these

6. _____ Would increase thirst
   a. Stimulating the medullary osmoreceptors
   b. Stimulating atrial volume receptors
   c. ANP
   d. All of these
   e. Two of these

7. _____ The macula densa is thought to communicate with the afferent arteriole via
    a. Paracrine secretions   d. ADH
    b. The sympathetic nervous system   e. None of these
    c. The parasympathetic nervous system

8. _____ People with diabetes mellitus
    a. May have a glucosuria
    b. Likely reabsorb more glucose than a normal person
    c. Do not produce ADH
    d. All of these
    e. Two of these

9. _____ If you could give a drug that blocked the reabsorption of potassium,
    a   The renal clearance of potassium would subsequently be greater than a simultaneous inulin measurement.
    b. Urine production would likely increase
    c. The filtration fraction would increase
    d. All of these
    e. Two of these

10. _____ PAH
    a. Is in equal concentration in the afferent and efferent arterioles
    b. Is used to measure the GFR
    c. Is in lower concentration in the renal vein than in the renal artery
    d. All of these
    e. Two of these

11. _____ As a consequence of respiratory acidosis...in an attempt to compensate
    a. One would stop filtering bicarbonate
    b. One would breathe more frequently
    c. One would secrete hydrogen ions into the filtrate
    d. All of these
    e. Two of these

For questions 12–18, choose one of the following:
    A. Increases or is greater than
    B. Decreases or is less than
    C. Has no effect or is equal to

12. _____ The concentration of sodium in the afferent arteriole, as compared with the concentration of sodium in the efferent arteriole

13. _____ The amount of PAH in the afferent arteriole, as compared to the amount of PAH in the efferent arteriole

14. _____ Assuming the same plasma concentration in the afferent arteriole, compare the amount of PAH in the efferent arteriole as compared to the amount of inulin in the efferent arteriole.

15. _____ The effect of increasing the filtration fraction from 20% to 30%, on the concentration of glucose in the efferent arteriole

16. _____ The tubular load of glucose in a healthy person, as compared to a diabetic

17. _____ The renal plasma threshold of glucose in a healthy person, as compared to a diabetic

18. _____ The effect of angiotensin II on the subsequent diameter of the efferent arteriole

CHAPTER 10

Ventilation

Introduction to Human Breathing and Controls

For clarity, we will first explain the anatomically correct way in which things work concerning ventilation/respiration. We will then define some terms the way we are going to use them.

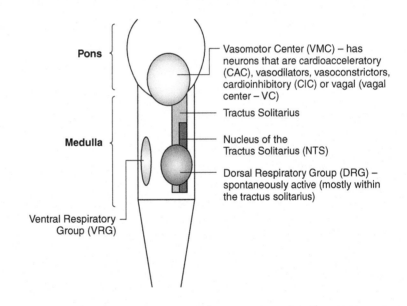

Pons

Vasomotor Center (VMC) – has neurons that are cardioacceleratory (CAC), vasodilators, vasoconstrictors, cardioinhibitory (CIC) or vagal (vagal center – VC)

Tractus Solitarius

Medulla

Nucleus of the Tractus Solitarius (NTS)

Dorsal Respiratory Group (DRG) – spontaneously active (mostly within the tractus solitarius)

Ventral Respiratory Group (VRG)

FIGURE 10.1

Dorsal Respiratory Group (DRG):   the neurons that make up the DRG are mostly located within the nucleus of the Tractus Solitarius of the medulla. These neurons are spontaneously active. That is to say that if an animal has had everything else removed, these neurons will still depolarize regularly. They send signals via the phrenic nerve to the diaphragm (causing it to contract) and via spinal nerves to the external intercostals (causing them to contract), both of which increase the thoracic volume (inspiration).

Nucleus of the Tractus Solitarius (NTS): This is the part of the Tractus Solitarius within the medulla in which sensory information from baroreceptors, chemoreceptors, and stretch receptors from the lungs terminate. The NTS then can do many things once it has received these signals. The NTS may send signals to the VMC to do a number of things (vasodilate, vasoconstrict, increase HR, decrease HR). Also, when the NTS receives feedback from stretch receptors, it shuts off the DRG, stopping inspiration.

Vasomotor Center (VMC): This is located bilaterally in the medulla and the lower half of the pons. Mostly within reticular formation, this center of neurons houses many types of neurons. It contains a subset of vagal neurons, which will send information via the vagus nerve to decrease HR. It also contains cardio-acceleratory center (CAC) neurons, which will send signals via spinal nerves to increase HR. This center also contains groups of neurons that can cause vasodilation or vasoconstriction.

VRG: the group of neurons responsible for forced expiration.

In order to better understand the relationship between these centers, it is necessary that we operationally define some terms:

VMC/CAC (vasomotor center/cardio-acceleratory center): We will use this term to mean neurons within the VMC that act to vasoconstrict and increase HR.

CIC/VC (cardio-inhibitory center/vagal center): We will use this term to mean the neurons (some within the VMC) that act to decrease HR via a vagal influence.

<u>**Important interactions**</u>

VMC/CAC has a facilitatory role on the DRG (increase VMC/CAC, and you increase breathing).

Baroreceptors send signals to the NTS to cause:
- decreased VMC/CAC
- increased CIC/VC
- subsequent decreased DRG activity (because of VMC/CAC decrease)

Chemoreceptors send signals to the NTS to cause:
- increased VMC/CAC
- decreased CIC/VC
- subsequent increased DRG activity (because of VMC/CAC increase)

Signals from the lungs (stretch receptors) are sent via the vagus to the NTS to cause:
- decrease or "turn off" of DRG activity (Herring Breuer reflex).

Signals from the arch of the aorta are sent via the vagus (X) nerve to the NTS.

Signals from the bifurcation of the carotids are sent via the glossopharyngeal (cranial nerve IX) nerve to the NTS.

Although we will see that CO_2 drives respiration, it is important to note that peripheral chemoreceptors are much more sensitive to changes in CO_2, while the central chemoreceptors (sensitive to H^+) are much less sensitive to CO_2 changes. However, central chemoreceptors are sensitive to plasma CO_2 changes within the physiological realm of normal breathing (PCO_2: 35–75 mm Hg). **Thus, CO_2 drives respiration at the level that is mediated within the medulla chemosensitive area** (perhaps peripheral chemoreceptors have a role in exercise though, as they are more sensitive to PCO_2 changes).

I. **Regulation**
   a. **Establishment of Basic Pattern:** Why do we breathe? We need oxygen to make water (oxygen is the final electron acceptor in the electron transport chain in aerobic metabolism). We also breathe to get rid of CO_2. Regulated by neuronal mechanisms (mainly located in the medulla—some are in the pons). Breathing is a spontaneous activity (we could remove your cortex and leave your brainstem, and you would continue to breathe—albeit you would not be aware that you were breathing).
   i. **Medullary Structures**
      (1) **Dorsal Respiratory Group** (mostly within the NTS). "Inspiratory center" (remember, "center" = a number of neurons associated together to perform a specific function)

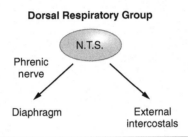

FIGURE 10.2

   a. DRG is spontaneously active. Impulses sent via the phrenic nerve to the diaphragm cause it to contract (pull inferiorly). This expands the volume of the thoracic cavity. The DRG also sends impulses via the spinal nerves to the external intercostals (between the ribs), causing them to contract and expand the volume of the thoracic cavity.
      (2) **Ventral Respiratory Group** (medullary center ventrally located): "Expiratory center"

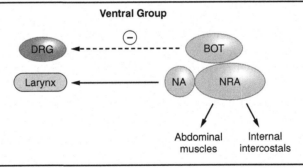

FIGURE 10.3

   a. **Nucleus ambiguous:** Dilates the larynx and the bronchioles (airways).
   b. **Nucleus retro-ambigualis:** Causes the abdominal muscles and the internal intercostals to contract (decreasing the volume of the thoracic cavity). **This is forced expiration—not normal expiration!**
   c. **Botzinger's complex:** Inhibits the DRG (inspiration) and allows expiration.

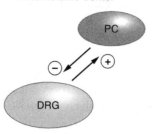

Pneumotaxic Center

PC

DRG

FIGURE 10.4

ii. **Pontine Structures:**
  (1) **Pneumotaxic Center (PC):** When the DRG is excited (spontaneously) it also sends impulses to the PC, which in turn shuts off the DRG. The respiratory pattern is established this way.
    a. We breathe in because we actively contract the external intercostals and the diaphragm. However, **exhaling in normal situations is a passive process!** The DRG "shuts off" to exhale.
  (2) **Apneustic Center (AC):** Role is not 100% clear. We *do* know it does something to **amplify inhalation**. Stimulation of the AC alters the respiration pattern: causes one to breathe in deeply (and for a prolonged period), with quick bursts of expiration. **It amplifies inspiration.**
b. **Neuronal Control**
  i. **Other parts of CNS**
    (1) **Brain**
      a. **Cerebrum:** Capable of influencing the basic breathing pattern. Holding our breath interrupts the basic pattern of breathing (controlled by the cerebrum). Eventually one would pass out because the cerebrum is not getting enough oxygen. At this point you cannot consciously hold your breath any longer and normal respiration would resume. The speech centers (Broca's) also alter the basic patterns of respiration (we learn to talk and breathe at the same time).
      b. **Hypothalamus:** Alters the basic pattern for some animals. Dogs pant to cool themselves by evaporative cooling of structures within the mouth and pharynx. **Does this cause a respiratory alkalosis in the dog? Why or why not?**

      c. **Limbic system:** Emotions such as **anxiety and fear** were found to control respiration (increase it).
    (2) **Spinal Cord:** A facilitative ("cord facilitation") effect on the basic pattern. Afferent pathways that run in the spinal cord to the medulla spill over their electrical activity into respiratory centers. If we get doused with ice cold water, there is a tremendous amount of afferent impulses going up the spinal cord. The electrical activity spills over into the respiratory centers and causes us to have a response until the cerebrum tells us to calm down. Painful stimuli also can do this (an example is spanking newborns to make them breathe).
  ii. **Stretch Receptors in Lungs (Herring Breuer [HB] reflex):** A reflex to keep the lungs from overinflating. Receptors in the bronchioles are sensitive to being stretched. When the receptors are excited (from stretch) they send impulses via the vagus nerve to the NTS to inhibit the DRG (stop inspiration).

   (1) Infants use this reflex to shut off inspiration, but **adults do not normally use this reflex (some lower animals *do* rely on this mechanism).**

   (2) If the vagus nerve is stimulated, respiration will decrease almost to the point of stopping (afferent fibers are telling it that the lungs are overinflating). **If the vagus nerve is cut in an animal that relies on the HB reflex to stop inspiration, then respiration will become deeper, with a slower rate.**

c. Humoral Control: The respiratory system has an important role in regulating CO_2, O_2, and H^+ ions in the arterioles. With increasing CO_2, increasing H^+, and decreasing O_2 in the arterioles, there is a positive effect on breathing rate (ventilation). PCO_2 causes the greatest effect. O_2, H^+, and CO_2 are linked together, but **CO_2 drives respiration (mainly through the centrally mediated CO_2 response).**

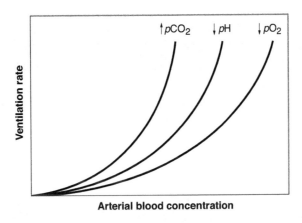

FIGURE 10.5

  i. **Role of arterial O_2:** Has the least effect on respiration in normal situations. Low O_2 is monitored by peripheral chemoreceptors at the arch of the aorta and at the bifurcation of the carotid arteries (sensitive to low O_2). pO_2 has to fall really low for the chemoreceptors to be affected (has to fall below a partial pressure of 60 mm Hg; partial pressure of arterial pO_2 at sea level is 100–105 mm Hg).

    (1) When excited, chemoreceptors send signals to the NTS, which tell the DRG to breathe **(at high altitudes the partial pressure of O_2 is low, so we may use this mechanism).**

  ii. **Role of arterial CO_2: Drives respiration.** The same chemoreceptors are also affected by CO_2, but only 30% of the CO_2 response is mediated by these peripheral receptors. There are other chemoreceptors on the medulla itself (a chemosensitive area) that are responsible for the bulk of the response to elevated CO_2.

    (1) Center that *drives* respiration is actually stimulated by H^+ ions! These receptors on the medulla drive respiration by being sensitive to high H^+ ion concentrations in the CSF (which result from high arterial CO_2).

    (2) Remember the blood-brain barrier: If there is a high H^+ ion concentration in the blood, it affects the peripheral chemoreceptors only because H^+ ions cannot get across the blood-brain barrier.

    (3) CO_2 goes across the blood-brain barrier, but the receptors on the medulla are sensitive only to H^+ ions! The enzyme carbonic anhydrase catalyzes

$CO_2 + H_2O \rightarrow H_2CO_3 \rightarrow H^+ + HCO_3^-$. Subsequently, H^+ stimulates the medulla. **By this indirect mechanism, plasma CO_2 has the most profound effect**.

    iii. **Role of arterial H^+**: H^+ ions in the blood also have an effect on the ventilation rate. The same chemoreceptors at the arch of the aorta and the bifurcation of the carotid arteries respond to elevated H^+ levels. Elevated H^+ ions cause receptors to get more excited and increase ventilation/respiration.

  d. **Effect of Exercise:** Exercise increases the ventilation rate (15- to 20-fold).
    i. **Heavy exercise:** Ventilation keeps pace with the demand.
    ii. **H^+, CO_2, O_2** —these hardly change in people who exercise.
    iii. **One proposal is that ventilation changes by moving the appendages** (receptors are sensitive to movement in the joints and muscles). An experiment was done in a person who was sitting still but who was passively having his appendages moved. This affected the normal respiration rate and increased body temperature. **Another proposal is that higher centers in the brain are responsible;** motor areas may stimulate the respiratory centers at the same time they initiate the movements (a "feed-forward" mechanism).

  e. Relationships between **vasomotor center and respiratory center:**

The Relationship between VMC/CAC and the DRG		
	Baroreceptors	**Chemoreceptors**
What stimulates?		
Effect on vasoactivity?		
Effect on HR?		
Effect on ventilation (DRG)?		

Complete the chart above in class *before* reading the following explanations.
    i. **The respiratory center is close to the VMC/CAC.**
      (1) The baroreceptors and the chemoreceptors are at the arch of the aorta and at the bifurcation of the carotid arteries (signals from the arch of the aorta travel via the vagus nerve [X], while signals from the bifurcation of the carotid arteries travel via the glossopharyngeal nerve [IX]).
    ii. **Stimulation of the baroreceptors** is due to stretching by high BP.
    iii. **Stimulation of the chemoreceptors** is due to an increase in the amounts of H^+ ions, CO_2, and/or a decrease in O_2.
    iv. **Vasoactivity when the baroreceptors are stimulated:** Leads to vasodilation via increased signals to NTS (which cause an increase in signals to inhibit the VMC/CAC and excite the CIC/VC).
    v. **Vasoactivity when the chemoreceptors are stimulated:** Causes vasoconstriction via signals to NTS that excite the VMC.
    vi. **Heart rate decreases due to baroreceptor stimulation:** Increased signals to the NTS (which cause an increase in signals to inhibit the VMC/CAC and excite the CIC).
    vii. **Heart rate increases due to chemoreceptor stimulation:** Increased chemoreceptor signals to the NTS cause less inhibition of CAC and less excitation of CIC/VC.
    viii. **Ventilation:**
      (1) Ventilation increases when the chemoreceptors are stimulated. This is because the VMC/CAC has a facilitative effect on the respiratory centers (i.e., when the VMC/CAC increases in activity, the respiratory centers also increase in activity).
      (2) Ventilation when the baroreceptors are stimulated = temporary apnea (a short period of time when breathing stops). This happens because of the aforementioned

relationship between the VMC and CAC and the respiratory center. When baroreceptors are stimulated, you get more inhibition of the VMC and CAC; thus you also get less facilitation of the respiratory centers (i.e., breathing decreases).

Given what we have just learned, what happens to the ventilation of an animal if we inject epinephrine? (Think it through, from the effects on BP to the effects on the respiratory centers)

II. Anatomy of Respiratory System

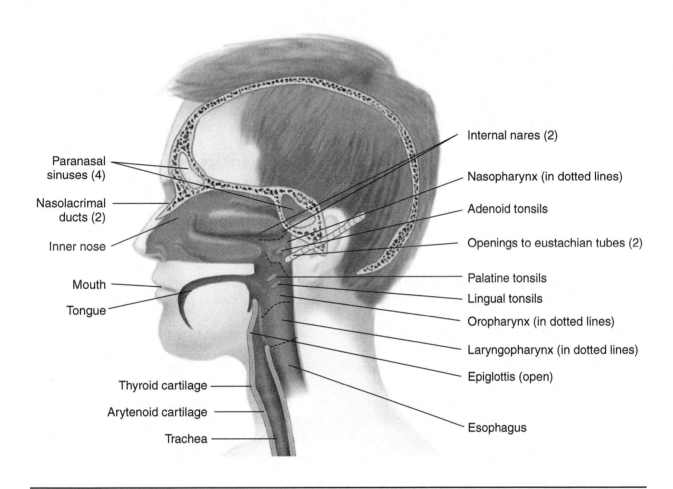

Paranasal sinuses (4)

Nasolacrimal ducts (2)

Inner nose

Mouth

Tongue

Thyroid cartilage

Arytenoid cartilage

Trachea

Internal nares (2)

Nasopharynx (in dotted lines)

Adenoid tonsils

Openings to eustachian tubes (2)

Palatine tonsils

Lingual tonsils

Oropharynx (in dotted lines)

Laryngopharynx (in dotted lines)

Epiglottis (open)

Esophagus

FIGURE10.6

a. **Nose**
   i. **External portion:** External nares lead to internal nose.
   ii. **Internal portion:** In the skull; extends back and merges with the pharynx. Leads to the pharynx through the **internal nares**. There are four paranasal sinuses, which communicate with the internal nose. These play a role in sound and quality of voice. There are also **two nasolacrimal ducts** (these come from the eyes to the internal nose and help drain tears from the eyes—they are working all the time. Under emotional duress, the capacity to drain tears may be exceeded and the excess tears run out of the eyes).

iii. **Functions**
   (1) **Olfaction:** The ability to smell via phasic bipolar neurons
   (2) **Condition incoming air:** Warm, moisturize, and cleanse it
   (3) The nose and the paranasal sinuses are hollow and resonating chambers that help to **add resonance and quality/tone to the voice.**
b. **Pharynx:** Three parts; starts with the internal nares and runs down the neck.
   i. Nasopharynx: four openings:
      (1) **Two internal nares**
      (2) **Two openings to Eustachian tubes:** Connects the pharynx to the middle ear; helps to equilibrate the pressure on both sides of the eardrum.
      (3) Along the dorsal wall are the **pharyngeal tonsils (adenoid tonsils).**
   ii. Oropharynx: The middle portion
      (1) **The fauces:** Opening between the mouth and the oropharynx
      (2) Two pairs of tonsils: **palatine tonsils** and **lingual tonsils**
   iii. **Laryngopharynx:** Empties the contents into the esophagus or the larynx. This has both respiratory function and digestive function.
c. **Larynx:** Short passageway that connects the pharynx to the trachea; supported by **nine pieces of cartilage**
   i. **There are three pairs of cartilage and three single cartilage pieces.**
   ii. **Arytenoid cartilage:** Paired and positioned along the anterior wall of the larynx. These attach to and adjust the tension of the vocal chords. These can alter the stretch of the vocal chords and subsequently change the pitch from low to high.
   iii. **Thyroid cartilage:** A single cartilage on the anterior wall of the larynx. The size is influenced by androgens (mainly testosterone). Thus it is larger in males than in females (known as the Adam's apple).
   iv. **Epiglottic cartilage (epiglottis):** A "trap door" that covers the opening into the trachea when we swallow food. The epiglottis flops down so that food is propelled into the esophagus and not into the trachea.
d. **Trachea (windpipe):** A tubular structure 12 cm in length that extends to the 4th–5th thoracic vertebra.
   i. Branches into the **right primary bronchus** and the **left primary bronchus.**
   ii. **Carina:** A ridgelike structure; the most sensitive of the respiratory structures. If foreign material comes in contact with the carina, it induces coughing in order to eject the material.
   iii. The trachea has a number of cartilaginous rings (16–20) that are shaped like a C. The open part of the C is pointing towards the back (this is where the esophagus would sit). The trachea is a cartilaginous structure, so it will not collapse upon lowered pressures (compare this to the hose of a vacuum cleaner).
      (1) There are **goblet cells (these contain cilia)** in the trachea that generate mucus.
   iv. The right bronchus has a greater diameter than the left bronchus and is more directly parallel to the trachea. As a result, larger inhaled foreign material more commonly becomes lodged in the right lung.
e. **Bronchi:** The primary bronchi divide into secondary bronchi (or lobar bronchi), then into tertiary bronchi (or segmental bronchi), then into more divisions until they ultimately lead into terminal bronchioles, which lead to the respiratory units.
   i. Everything before the respiratory unit is part of the respiratory system but does not participate in gas exchange. Thus, these structures make up what is known as the **dead space.**
      (1) The dead space's function is to humidify and clean the air before it reaches the respiratory unit.

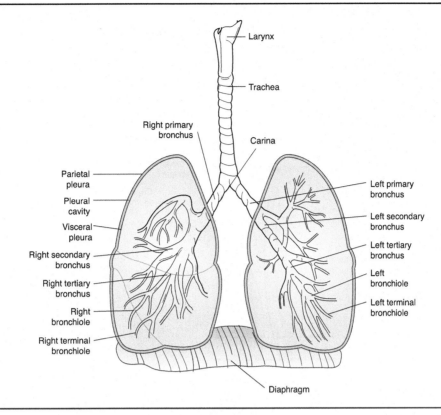

FIGURE 10.7

f. **Respiratory Unit:** Where gaseous exchange takes place. The 300 million alveoli supply around 750 square feet of surface area for gas exchange.

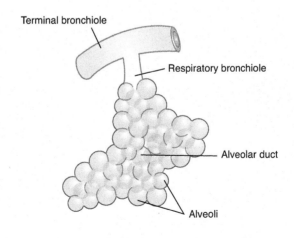

FIGURE 10.8

g. **Respiratory Membrane:** The surface of the epithelium (inside the alveolus) and the respiratory capillary; the surface of the epithelium is covered by water and surfactant.
  i.   There is a very short diffusion distance (shown below). Anything that causes edema in the lungs affects this distance and subsequently compromises lung function.
  ii. **Congestive heart failure:** BP in the capillaries of the alveolus increases and more fluid is filtered into the interstitial space. This resultant edema now makes the diffusion distance wider, which interferes with gas exchange.

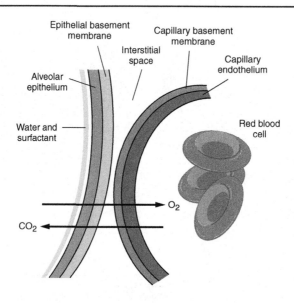

FIGURE 10.9

iii. **There are different cell types:**

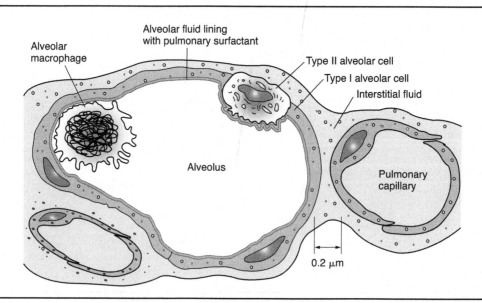

FIGURE 10.10

      (1) **Type I alveolar cells:** Most of the alveolar cells; the basic structural cells

      (2) **Type II alveolar cells:** Produce pulmonary surfactant

      (3) There are also macrophages in the alveolus.

  h. **Surfactant:** This substance decreases the surface tension of the water at the epithelial membrane.

    i. When you breathe in, the alveoli (lined with water and surfactant) must stretch. To do this, you must overcome the surface tension of the water inside of the alveolus. Surfactant lowers this surface tension and makes it more feasible to do so.

    ii. Surfactant also has a role in alveolar stability:

      (1) There is a relationship between the pressure in the sphere (alveolus) and the surface tension.

      (2) $P = (2T)/R$

      (This formula explains why a bike tire with a small radius can withstand a higher pressure for the same wall tension as a larger car tire.)

          T = surface tension

          R = radius of alveolus

          P = pressure

      (3) As R gets smaller, the pressure increases. Thus, the smaller alveolus would collapse into the bigger alveolus.

      (4) **Surfactant prevents this by creating more stability in the alveoli (decreases surface tension).**

 i. **IRDS & ARDS**

    i. **IRDS (infant respiratory distress syndrome):** In general, surfactant appears late in a fetus's life because the infant does not need to breathe until it is born (pulmonary structures and fetal surfactant are fully functional around 34 weeks).

      (1) **In premature infants** (or with a mother who has diabetes mellitus or other pathology), there is not sufficient surfactant produced and the infant has a great deal of trouble breathing. This can lead to death.

      (2) **Treatment:** Artificial surfactant can be administered through a tube or a breathing apparatus can be put in to force air in the lungs.

    ii. **ARDS (adult respiratory distress syndrome):** There is a different etiology involved than with IRDS. Also known as "wet lung" or "shock lungs," 50% of the people with ARDS die even with medical treatment. The exact cause is not known, but thus far it is believed that it is due to some disturbance in the pulmonary capillaries:

      (1) **The pulmonary capillaries become very permeable to fluid** (because of near-drownings [water in the lungs], reactions to toxins, certain surgeries [heart bypass], or infections from pneumonia or TB).

      (2) The resultant leakage of proteins into interstitium causes a **pulmonary edema, which** forces water and fluid into the alveoli. This destroys the surfactant (notice the ARDS patients *do* produce surfactant).

III. Mechanics of Ventilation: How do we get air in and out of the lungs?

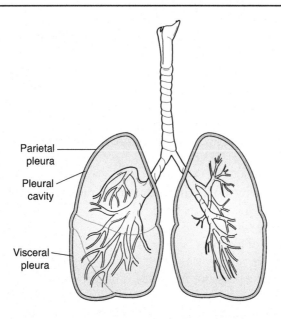

Parietal pleura

Pleural cavity

Visceral pleura

FIGURE 10.11

a. **Pleural membranes**
   i. **Visceral pleura:** A pleural membrane that is "plastic wrap–like" on the outside of the lungs; this actually lines the lungs.
   ii. **Parietal pleura:** Lines the walls of the thoracic cavity
   iii. Between these membranes is the **pleural cavity (intrapleural space, interpleural space),** which is mainly a potential space. Some fluid is there to help lubricate the two layers.
      (1) Pleurisy: An inflammation of one or both of the membranes
b. **Pressure differences and dynamics: In order for air to enter the lungs, intra-alveolar pressure must be *less* than the atmospheric pressure. In order for air to go out of the lungs, intra-alveolar pressure must be *more* than atmospheric pressure.**
   i. Expanding the alveolar wall will decrease pressure so the air will come in.
   ii. In order for the lungs to expand and contract, a negative pressure system must exist. **The lungs inflate by changing the pressure around the lungs (intrapleural pressure).**
   iii. If there is no difference in the pressure, then air is not moving in either direction.
      (1) If atmospheric pressure is 760 mm Hg and intra-alveolar pressure is 760 mm Hg (because the epiglottis is open), then the air in both places is equal and thus not moving in either direction.
   iv. We need some pressure gradient so that the alveoli will inflate.
   v. Impulses from the DRG cause the volume of the thoracic cavity to expand (diaphragm pulls inferiorly and external intercostals pull superiorly). The parietal pleura starts pulling away from the visceral pleura and thus increases the pleural cavity space. This decreases the pressure in the pleural cavity and causes the visceral pleura to move out toward the parietal pleura. This creates a negative pressure in the alveoli.
      (1) The alveoli expand (air goes into the lungs from the atmosphere).

(2) **If the DRG is inhibited, then the muscles relax. Thus the intra-pleural space decreases (which causes the pressure in this cavity to increase up to 758 mm Hg). Because of surface tension and elastic recoil, this is sufficient pressure to expel air from the lungs (i.e., an intrapleural pressure of 758 mm Hg is enough to generate an intra-alveolar pressure that is greater than 760 mm Hg and thus causes expiration). Normally, when the intra-alveolar pressure is equal to the atmosphere (760 mm Hg), we maintain an intrapleural pressure of around 755 mm Hg to keep the alveoli from being collapsed.**

vi. If there is **no pulmonary surfactant**, the 752 mm Hg pressure usually seen in the intra-pleural space during inspiration would change the most in order to overcome the increased surface tension.

vii. **Pneumothorax:** Opening from intrapleural space to atmosphere.
From what you know, what happens to the lung in this instance? Would one or both lungs collapse?

viii. When alveoli collapse it is called **atelectasis**.

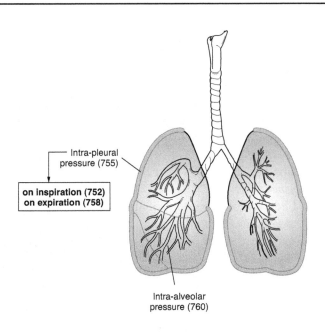

Intra-pleural pressure (755)

on inspiration (752)
on expiration (758)

Intra-alveolar pressure (760)

FIGURE 10.12

IV. Respiratory Volumes: Measured with a spirometer (measures the difference in volumes).

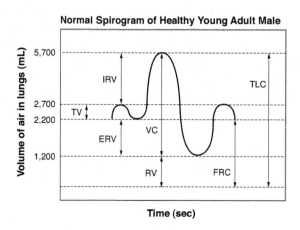

Normal Spirogram of Healthy Young Adult Male

FIGURE 10.13

Refer to the above chart to get an idea of what each term is.
   a. **Tidal Volume (TV):** The normal breathing (in and out).
      Think: the tides come in and out like your breath.
   b. **Inspiratory Reserve Volume (IRV):** Breathing in beyond normal as deeply as possible.
      The "reserve" beyond a normal inspiration.
   c. **Expiratory Reserve Volume (ERV):** Breathing out beyond normal as much as possible.
      The "reserve" beyond a normal exhalation (that can be breathed).
   d. **Residual Volume (RV):** Lung volume that is never breathed out.
      This volume always "resides" in lungs.
   e. **Functional Residual Capacity (FRC):** RV + ERV
      After a functional breath, what is left?
   f. **Vital Capacity (VC):** The maximum amount of air that can possibly be exchanged.
      At your most "vital," how much air can you move?
   g. **$FEV_{1.0}$:** How much one can breathe out in one second. (Breathe in deeply and then blow out.)

V. Airway Resistance and Pathologies
   a. Normal
      i. Airflow $= \Delta P/R$
      ii. $\Delta P=$ Difference between atmospheric pressure and intra-alveolar pressure.
      iii. R= Resistance of the airways…determined by the radius.
      iv. In normal persons the radius is large enough so that the resistance is low. As a result, only slight differences in pressure are necessary.
      v. In normal individuals the autonomic nervous system is capable of sufficiently altering the diameter of the bronchi to meet the various demands for oxygen.
      vi. There are pathological situations that can impair the flow of air.
   b. Pathologies—Chronic Obstructive Pulmonary Disease (COPD): A group of lung diseases characterized by a decrease in the diameter of the bronchi causing an increase in resistance. This, in turn, causes a significant increase in the gradient (the difference between atmospheric pressure

and intra-alveolar pressure) necessary to get adequate air into and out of the lungs. In COPD expiration is more of a problem than inspiration.

    i. **Chronic Bronchitis**—Chronic inflammation and accumulation of mucus in the airway caused by exposure to various irritants such as cigarette smoke or pollutants in the air.

    ii. **Asthma**—Thickening of the walls of the airway caused by inflammation and the release of histamines. Oversensitivity of the airway causes the bronchi to constrict. Often caused by allergens.

    iii. **Emphysema**—Breakdown of alveolar walls due to the release of enzymes such as trypsin from macrophages in response to irritants such as cigarette smoke.

  c. Pulmonary Fibrosis—not COPD; rather, normal lung tissue is replaced by connective tissue. Causes the lung to be inelastic or make it less compliant.

VI. Gaseous Exchange: In order to change the composition of any fluid compartment in the body, you have to start at the plasma. Gas exchange takes place at the **pulmonary capillaries** and the **systemic capillaries.**

  a. **Physical Laws**

    i. **Dalton's Law (The Law of Partial Pressures):** Each gas within a mixture of gases contributes to the total pressure. **The partial pressure exerted by a gas is proportional to its mole fraction times the total pressure.**

      (1) 760 mm Hg $\times$ 0.21 O_2 = 160 mm Hg of pO_2. The partial pressure of O_2 (pO_2) in the air is equal to 160 mm Hg.

      (2) CO_2 is important, too: $760 \times 0.0003 = 0.23$ mm Hg of pCO_2 in the atmosphere.

      (3) Partial pressures are important because the gases will always move from a higher partial pressure to a lower partial pressure.

    ii. **Henry's Law: The ability of gases to become dissolved in liquids.** Two things affect this:

      (1) Partial pressure of the gas over a liquid (all other things equal)

      (2) Solubility of the gas in the liquid. There are different solubility coefficients. CO_2 is 20 times more soluble in water than O_2.

    iii. **Hyperbaric oxygenation** (greater than atmospheric oxygen): Atmosphere with three to four times the concentration of O_2 than normal. This may be used for treating infections such as anaerobic bacteria (tetanus or gangrene). Carrying more O_2 in the blood and tissues is toxic to anaerobic bacteria.

    iv. **Nitrogen** is not a very soluble gas at 760 mm Hg.

      (1) Scuba divers: Because they descend to very high pressures, nitrogen does begin to dissolve into their blood. If they ascend too quickly, the nitrogen will bubble out in their blood vessels ("the bends"). These divers must ascend slowly, allowing time for the nitrogen to dissolve out and to be breathed off slowly.

    v. **pO_2 in the atmosphere is 160 mm Hg.** It comes in the nose and mouth to the pharynx, larynx, trachea, bronchi, and finally the alveoli.

      (1) As air comes into the body (into the dead space), we humidify the air. Water vapor is being added to the air, so the partial pressure of each gas decreases accordingly.

      (2) At the alveoli there is equilibration across the lungs with the blood.

      (3) Expiration partial pressure of O_2 is between "humidified air" and "equilibrated air."

    vi. **pCO_2 atmospheric partial pressure is 0.23 mm Hg.** Water vapor does not change this value a great deal.

      (1) The partial pressure will increase as it equilibrates with arterial blood (40 mm Hg) in the alveoli.

      (2) Expiration partial pressure of CO_2 is between "humidified air" and "equilibrated air."

	Atmospheric	Upper Airway	Alveoli	Expired
pO_2 mm Hg	160	149	104	120
pCO_2 mmHg	0.23	0.23	40	27

b. **Exchange:** Exchange takes place only at the capillaries (pulmonary and tissue).

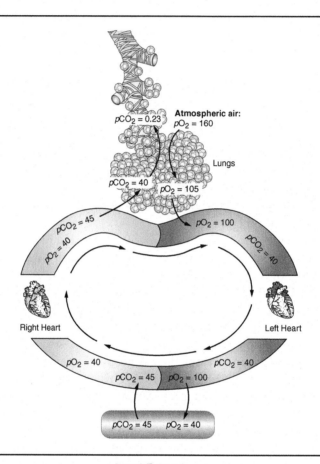

Atmospheric air:
$pO_2 = 160$

$pCO_2 = 0.23$

Lungs

$pCO_2 = 40$ $pO_2 = 105$

$pCO_2 = 45$ $pO_2 = 100$

$pCO_2 = 45$

$pO_2 = 40$

$pCO_2 = 40$

Right Heart   Left Heart

$pO_2 = 40$ $pCO_2 = 40$

$pCO_2 = 45$ $pO_2 = 100$

$pCO_2 = 45$ $pO_2 = 40$

FIGURE 10.14

i. **O_2:** Gases move down the concentration gradient. At the alveoli there is a very quick equilibration to a pO_2 of 100–105 mm Hg.
   (1) At the heart the pO_2 still equals 100–105 mm Hg.
   (2) At the systemic capillaries: **Mainly because of the drop in pO_2, the O_2 in the blood
   diffuses quickly across to the tissues.** This diffusion is so quick that the blood equili-
   brates with the pO_2 of around 40 mm Hg that is present in the interstitium. Other things
   do facilitate unloading at the tissue level (Bohr effect).
   (3) The blood then returns all the way to the pulmonary capillaries with a pO_2 of 40 mm Hg,
   at which point atmospheric O_2 will again cause the quick equilibration of pO_2 in the
   alveoli to 105 mm Hg.

ii. **CO_2:** pCO_2 in the alveolus is 40 mm Hg. The CO_2 is unloaded from the blood to the alveolus because the pressure is higher in the blood than in the alveolus or the atmosphere. Again, it is diffusion that works in response to partial pressure differences.

iii. **Factors influencing diffusion:** If the respiratory membrane surface area is decreased, it will impair the gas exchange (like with TB, cancer, or congestive heart failure).

(1) CO_2 is much more soluble than O_2.

VII. Gaseous Transport

a. Hemoglobin

i. We have hemoglobin (Hb) because O_2 is not very soluble in plasma. Only 0.3 mL of O_2 is dissolved per 1 dL of blood (versus 19.6 mL of O_2 in this same amount of blood, but bound to hemoglobin). Hemoglobin is a complicated molecule with 4 heme groups (a heme group is an iron with several pyrrole rings) and a protein with 4 globular subunits. Many of these are packaged into our RBCs.

(1) Hemoglobin has the ability to combine loosely and reversibly with O_2:

$$Hb + O_2 \leftrightarrow HbO_2$$

(2) Oxygenation (not oxidation)!

(3) If O_2 is at a low partial pressure, then we have free hemoglobin; but as the partial pressure of O_2 increases, then we have more and more oxyhemoglobin (HbO_2). View this relationship in light of the rule of mass action (more O_2 will drive the reaction to the right; as the pO_2 drops, there is a tendency to disassociate).

ii. **Oxyhemoglobin Dissociation Curve:** At any partial pressure there is some equilibrium that exists between hemoglobin and oxyhemoglobin.

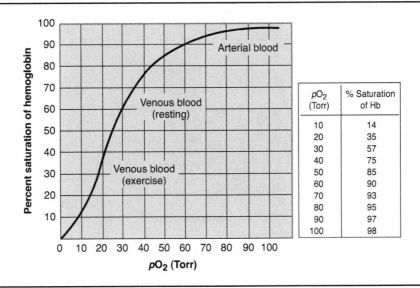

FIGURE 10.15

Note: Recall that "Torr" and "mm Hg" are equal (to be exact 1 Torr = $^1/_{760}$ of one atmosphere; one atmosphere was defined as 101,325 pascals (Pa) in 1954; mercury height can vary very slightly with temperature and gravity—you should think of Torr and mm Hg as equal).

(1) **% saturation of Hb as a function of pO_2**

(2) Observe that the **curve is sigmoidal** (due to the fact that once hemoglobin binds one O_2 molecule, it has an easier time binding the next one, etc.). At a pO_2 of 105 mm Hg, virtually all hemoglobin is saturated (97%–98%). Even if we decrease to a pO_2 of 80 mm Hg (mild elevation), there is no real change in saturation. However, at some point as we decrease the pO_2 we will "fall off" the steep portion of the curve and get rapid desaturation for given pO_2 decreases.

(3) **P_{50}** = the partial pressure at which we get 50% saturation (normally for hemoglobin this is about 30 mm Hg).

(4) Keep in mind that **the heart already extracts 75% of O_2**, so unlike other tissues that can increase O_2 extraction during intense metabolic demand, the heart must increase blood flow. Thus, coronary artery disease (CAD) is a big problem.

(5) **Factors affecting hemoglobin binding O_2:** Selection pressure would favor mechanisms that increase O_2 binding at the lung level and decrease O_2 binding at the tissue level. This is exactly what happens.

   a. **Bohr effect:** The presence of CO_2, H^+ ions (pH), or an increase in temperature will shift the oxyhemoglobin curve to the right (this allows for better unloading at the tissue level, where all of these are increased). These things all "disfavor" the binding of oxygen to hemoglobin. The **Bohr effect** *per se* refers specifically to the effect of CO_2 and pH on binding.

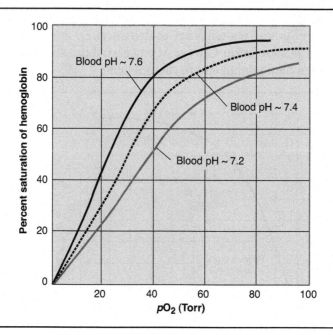

FIGURE 10.16

Maternal vs. Fetal Hemoglobin: Complete the chart below based on where you think a curve for <u>fetal hemoglobin</u> would be.

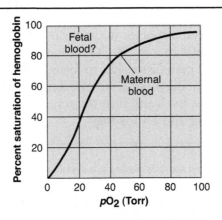

FIGURE 10.17

b. **Exposure to Altitude (Hypoxia): Is there a lower or higher % of O_2 at altitude?**

Upon going to elevation, you will acclimatize by making more 2,3 DPG. This shifts the curve to the right. This is advantageous for most elevations. **Look at the curve again—at what altitudes would this build up of 2,3 DPG and subsequent shift of the curve to the right be disadvantageous?**

When giving CPR (and breathing your expired air into someone else's lungs), is their hemoglobin getting mostly saturated from the O_2? Explain.

Before we leave altitude: Based on what you know about chemoreceptors and baroreceptors, what would happen to breathing and subsequent blood gas levels upon your first 2 to 3 hours at high altitude? Explain.

c. Hemoglobin (Hb) does not exclusively bind oxygen. Other things bind to Hb with great affinity. **Carbon monoxide (CO)** binds to Hb with approximately 200 times the affinity of O_2 (and thus the CO curve is *not* sigmoidal, but rather more like a fetal curve for O_2).
d. Hemoglobin also has a great affinity for NO (nitric oxide—an endothelial-derived relaxing factor). NO is localized; it only goes cell to cell so as not to saturate hemoglobin. **Hb binds NO even better than CO.**

b. **Transport of CO_2:** Occurs three ways
   i. **In solution:** 7% of CO_2 is dissolved in the plasma.
   ii. **Carbaminohemoglobin:** 23% of CO_2 is transported in the RBCs in this form. CO_2 also binds to hemoglobin, but to the globular portion (O_2, CO, and NO all bind to the heme portion of

hemoglobin). CO_2 on hemoglobin interferes to a degree with the O_2 binding site (the Bohr effect). Also, the presence of O_2 interferes with the ability of CO_2 to bind to hemoglobin (the Haldane effect). **Thus, we have the Haldane effect at the lungs and the Bohr effect at the tissues.**

    iii. **Bicarbonate:** 70% of CO_2 is transported in the form of bicarbonate. CO_2 is transported from the plasma into the RBCs to form H_2CO_3 in the presence of the enzyme carbonic anhydrase (CA). Carbonic acid then dissociates to form HCO_3^- and H^+ ions. HCO_3^- (bicarbonate) is then transported out to the plasma and exchanged for Cl^- (the "chloride shift"). This reaction happens at the tissue level, and then the reverse happens at the lung level.

  c. **Integration of CO_2/O_2 Loading and Unloading**

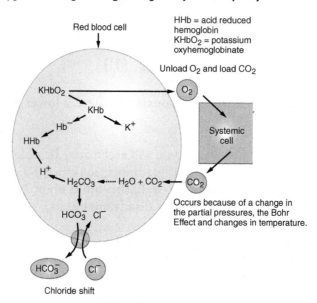

Oxygen unloading and CO₂ loading at a systemic capillary

FIGURE 10.18

Oxygen loading and CO₂ unloading at a pulmonary capillary

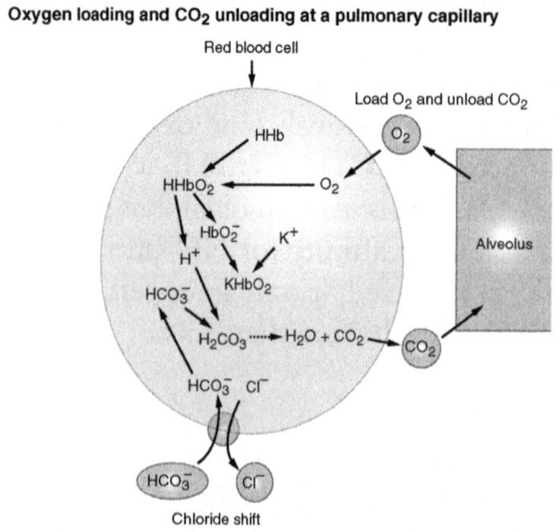

FIGURE 10.19

VIII. Red Blood Cell Production

a. **Components of Blood**

i. **There are two components:** Plasma and formed elements

ii. **Hematocrit** is the % of total blood volume due to the presence of the formed elements.

iii. The most predominant formed element is the **RBC**.

iv. Plasma has a number of inorganic and organic compounds.

(1) Organic: Proteins

a. Albumin: the most plentiful

b. Globulins: alpha, beta, and gamma

c. Fibrinogen: important in clotting factors

d. **All are made in the liver, except the gamma globulins.**

(2) Inorganic: Water, gases, electrolytes

v. **Formed Elements**

(1) **Erythrocytes: 4.6–6.2 million/mm³ of blood.** RBCs are the most prevalent component (>99% of the formed elements); about 25–30 trillion RBCs in the body

(2) **Leukocytes: 5–10 thousand/mm³ of blood.** There are two types.

a. **Granulocytes:** Have cytosolic granules (these stain according to the pH of the reagent).

- **Neutrophils:** Most prevalent formed element besides RBCs (stains in neutral solution)

- **Eosinophils** (stains in acid solution)

- **Basophils** (stains in basic solution)

b. **Agranulocytes:** No cytosolic granules

- **Lymphocytes:** Most predominant agranulocyte

- **Monocyte**s

(3) **Thrombocytes (platelets): 0.2–0.4 million/mm³ blood.** Fragments of megakaryocytes; important role in hemostasis

b. **Erythropoiesis:** RBC production. RBCs live for only about 120 days.

i. We form about 2 million RBCs per second. Men tend to have more than women.

ii. **Where erythropoiesis takes place** depends on the age of the patient.

(1) **Fetus**

a. 1st Trimester: occurs in the yolk sac

b. 2nd Trimester: occurs in the liver and spleen

c. Final Trimester: occurs in the bone marrow

(2) **Child:** Bones are filled with red bone marrow and thus are capable of erythropoiesis.

(3) **Adult:** Red bone marrow is replaced with yellow bone marrow (which is not capable of erythropoiesis) in the long bones. Flat bones of the body still house red bone marrow (sternum, hip, etc.). Adults make RBCs mostly in these bones.

iii. **Steps in the formation of formed elements:** Hemocytoblasts are multipotential cells that can develop to virtually any formed element.

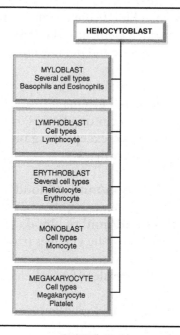

FIGURE 10.20

(1) **White blood cells:** Monoblasts, myloblasts, and lymphoblasts
(2) **Megakarocytes:** Parts of these break off and become **platelets.**
(3) **Erythroblasts:** These go through a number of cell divisions.
    a. **Erythroblast:** Has a nucleus, but is devoid of hemoglobin
    b. **Reticulocyte:** In the process of getting hemoglobin and losing the nucleus
    c. **Erythrocyte:** Has hemoglobin but no nucleus (mature RBC)
c. **Regulation:** We have to regulate the number of RBCs.

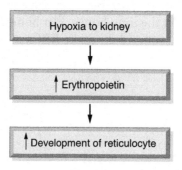

FIGURE 10.21

    i. **Hypoxia to the kidney** (too little oxygen to the kidney) causes erythropoietin to be produced. Erythropoietin in turn increases the rate of development of reticulocytes into erythrocytes. And thus hematocrit increases and we can carry more O_2 in the blood (as hematocrit increases, viscosity of blood also increases).

    ii. **Blood doping:** An illegal practice of endurance athletes. This is done by removing blood and storing it for several months before a competition. Then, prior to competition (and after the athlete's body has built its RBC count back up to normal), the athlete infuses the RBCs back into his or her blood. Thus, they now have a marked advantage in their ability to carry O_2 (and consequently a marked disadvantage in their ability to tolerate heat, as they have effectively reduced plasma and now are more susceptible to dehydration complications).

        (1) **(EPOGEN):** Drug that is a synthetic erythropoietin (EPO) (also illegal for athletes)

            a. Epogen was originally developed for patients (not athletes!) with **renal anemia**.

e. **Pathologies**

    i. **Anemia:** Too few RBCs or too little hemoglobin

        (1) **Nutritional:** Due to a deficiency in the diet (usually an iron deficiency). You need iron in order to make hemoglobin, and thus this is impaired if iron consumption is low. The old cure was to eat liver.

        (2) **Pernicious:** Due to an inability to absorb Vitamin B12 because of a lack of intrinsic factor in the stomach. Vitamin B12 is needed to form the stroma of the RBC. Without this, we form big, fragile RBCs (macrocytes) that are stuffed with hemoglobin and rupture easily.

            b. Vitamin B12 has to be injected for these patients, since they are unable to absorb it.

        (3) **Aplastic:** Due to bone marrow not making enough RBCs (usually a result of certain toxins, such as benzene, arsenic, or radiation)

        (4) **Renal:** Due to dysfunction of the kidney itself. The erythropoietin mechanism itself is damaged (happens in kidney failure). Treat with the drug EPOGEN.

        (5) **Hemorrhagic:** Due to hemorrhage or trauma of some sort (blood loss). Many women suffer from this because of menstrual flow. **Men have a higher hematocrit than women because of this and the fact that testosterone causes an increased erythropoiesis.**

        (6) **Hemolytic:** Due to sickle cell anemia (or **any type of pathology that manifests itself in a fragile RBC that hemolyzes easily**).

    ii. **Polycythemia:** Too many RBCs.

        (1) **Primary (polycythemia vera):** Due to a tumor-like condition in the bone marrow that stimulates the production of RBCs. Hematocrits can be 70–80! People with this have very thick blood; this increases the work load of the heart.

        (2) **Secondary (physiological):** Due to compensation for the decreased oxygen from making more RBCs. Occurs upon acclimatization to high altitudes. Hematocrits are around 50. Kidneys are chronically hypoxic at high altitudes and thus the EPO production increases to compensate. Also, any type of **lung pathology** could cause a **secondary pathological polycythemia (create an environment of hypoxia and hypoxemia due to lung impairment).**

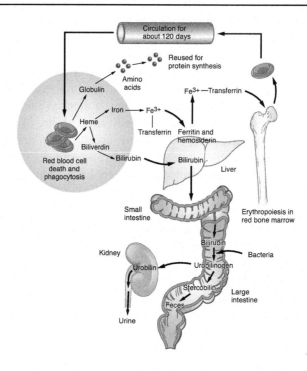

FIGURE 10.22

a. **Iron's fate:** When RBCs get older, they get more fragile. Eventually, as they go through the capillaries, especially at the spleen, they rupture.

   i. Iron is bound to a plasma protein, **transferrin**, which transports it to the liver. In the liver, iron is stored as **ferritin and hemosiderin.**

b. **Globulin's fate:** Globulins (1 protein with 4 subunits per hemoglobin molecule) are broken down into amino acids, which are then sent back to the general amino acid pool and used again later for protein synthesis.

c. **Pyrrole ring's fate:** The heme portion – iron = Pyrrole rings. The pyrrole ring is opened to a straight chain and converted to **biliverdin**. This is then converted to **bilirubin**, which is then transported to the liver. Bilirubin is taken up into the liver.

   i. **Bilirubin** is conjugated in the liver and finally **becomes part of the bile. This is then** sent to the gallbladder and eventually the small intestine (SI).

   ii. In the SI, **bacteria convert bilirubin into urobilinogen.** Urobilinogen can then be converted into **stercobilin** and excreted in the feces or absorbed as urobilinogen out of the SI and sent to the kidney. At the kidney, urobilinogen is made into **urobilin** and excreted in the urine.

   iii. What makes blood red? The heme in hemoglobin. Heme is a "respiratory pigment." Some animals have different-colored blood because they have different-colored respiratory pigments.
   The ice fish (inhabits very cold waters) has clear blood. What do you know about solution temperature and solubility that can help explain this?

(1) Jaundice: Condition in which the skin and whites of the eyes begin to turn yellow due to a buildup of bilirubin in the extracellular fluid. This can be due to a problem with the liver or the gallbladder (anything that causes a backup in bilirubin). This is a problem if it happens because bilirubin is neurotoxic. (Note: bilirubin is more toxic and less soluble than biliverdin, and yet we form it anyway because it is a very potent antioxidant). Infants not only usually have an immature liver at birth, but they also usually hemolyze many RBCs (fetuses have an elevated hematocrit to begin with). Because of this, usually an infant will become at least a little jaundiced (and very much so if premature). Thus, we put babies under UV lights (UV light destroys bilirubin, making it soluble).

    iv. **Stercobilin:** Responsible for the color of feces
    v. **Urobilin:** Responsible for the color of urine

X. Pharmacology
    a. **Nasal Decongestants:** Alpha-1 agonists. These act on the arterioles supplying the secretory glands of the nose.
        i. **Phenylephrine (NEOSYNEPHRINE)**
        ii. **Pseudoephedrine (ENTEX)**
    b. **Bronchodilators:** Treat asthma
        i. **(MAXAIR, VENTOLIN, PROVENTIL):** Beta-2 agonists
        ii. **(ATROVENT):** Anticholinergic (allows binding but blocks 2nd messenger)
        iii. **(SEREVENT):** Stimulates adenyl cyclase
        iv. **(SINGULAIR):** Inhibits leukotriene synthesis
    c. **Antihistamines:** H-1 antagonists (block histamine receptors). Not helpful for the common cold (although they may aid sleep).
        i. **(BENADRYL)**
        ii. **(CHLOR-TRIMETON)**
        iii. **(CLARITIN)**
        iv. **(ALLEGRA)**
    d. **Expectorants (guaifenesin):** Promote expectoration. These promote the ejection of sputum by making the sputum more watery (decreasing viscosity).
        i. guaifenesin and dextromethorphan **(GUAIFENESIN DM)**
        ii. guaifenesin and pseudoephedrine **(GUAIFENESIN PSE)**
    e. **Mucolytics:** Destroy mucus; used for people with **cystic fibrosis**
        i. **(MUCOMYST)**
    f. **Antitussives: Prevent coughing**
        i. Codeine: Antitussives work great but contain codeine and are highly addictive. They work because coughing is under CNS control at the medulla (the cough reflex). **Dextromethorphan: a noncodeine antitussive**

SUGGESTED READINGS

1. Guyton, A.C., and Hall, J.E. 2011. Chapters 37–41 in *Textbook of Medical Physiology*, 12th ed. Baltimore: Saunders.
2. Koeppen, B.M. and Stanton, B.A. 2008. Chapters 20-25 in *Berne & Levy Physiology*, 6th ed. St. Louis: Mosby.

CHAPTER 10 QUESTIONS

1. _____ Something you would be likely to find in the oropharynx
   A. Tonsils
   B. Openings into the Eustachian tubes
   C. Teeth
   D. All of these
   E. Two of these

2. _____ As a consequence of stimulating the stretch receptors in the Hering Breurer reflex, you would expect
   A. An increase in the number of afferent impulses in the 10th cranial nerve
   B. An increase in the number of efferent impulses in the phrenic nerve
   C. Inspiration to get deeper
   D. All of these
   E. Two of these

3. _____ At the level of systemic capillary, hemoglobin gives up oxygen because of
   A. The fall in partial pressure of oxygen
   B. The Haldane effect
   C. The increase in pCO_2
   D. All of these
   E. Two of these

4. _____ The form of hemoglobin in the pulmonary artery
   A. Acid-reduced hemoglobin
   B. Potassium oxyhemoglobinate
   C. Carboxyhemoglobinate
   D. Oxyhemoglobin
   E. None of these

5. _____ Contains a decongestant
   A. GUAIFENEX PSE
   B. NEOSYNEPHRINE
   C. SINGULAIR
   D. All of these
   E. Two of these

6. _____ Bilirubin
   A. Contains iron
   B. Is stored in the liver
   C. Its metabolite is found in the bile
   D. All of these
   E. Two of these

For questions 7–13, choose one of the following:
A. Increases or is greater than
B. Decreases or is less than
C. Has no effect or is equal to

7. _____ The production of surfactant in an infant suffering from IRDS, as compared to a healthy infant

8. _____ The magnitude of intrapleural pressure, as compared to the magnitude of intra-alveolar pressure, during expiration

9. _____ The magnitude of pulmonary compliance in a healthy person, as compared to someone suffering from pulmonary fibrosis

10. _____ The pO_2 of air if it moves from the atmosphere into the dead space

11. _____ The pO_2 in the renal artery, as compared to the pO_2 in the renal vein

12. _____ The p50 of maternal hemoglobin, as compared to the p50 of fetal hemoglobin

13. _____ The effect of increasing plasma H^+ on respiration in a person whose medullary chemo-receptors are not working

CHAPTER 11

Hepatic Physiology

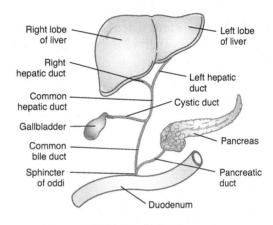

Right lobe of liver

Left lobe of liver

Right hepatic duct

Left hepatic duct

Common hepatic duct

Cystic duct

Gallbladder

Common bile duct

Pancreas

Sphincter of oddi

Pancreatic duct

Duodenum

Major Hepatic Structures

FIGURE 11.1

I. Introduction
   a. **General description & circulatory supply of the liver:** The largest metabolic organ in the body and the most important metabolically
      i. It **weighs 4 to 5 pounds** in an average adult.
      ii. It is divided into a smaller left lobe and a larger right lobe.
      iii. On the **inferior surface of the right lobe is the gallbladder (notice Figure 11.1 is drawn so that you can see the gallbladder—it is really tucked up under the right lobe of the liver).**
      iv. It is well vascularized and has a blood supply from two sources.
         (1) **Hepatic artery:** 25% of the liver's blood comes via this. This is just like any other artery in that it is a branch off of the aorta.
         (2) **Hepatic portal vein:** 75% of the liver's blood comes via this vein. Blood drains from the digestive system into the

hepatic portal vein. This vein then goes to the liver before returning to the heart. All blood leaves the liver to return to the heart via the **hepatic vein** (the hepatic portal system services everything from esophagus to large intestines).

b. **Functional Overview**
  i. **Metabolic Processes:** Major nutrients such as carbohydrates, fats, and proteins are processed in the liver after absorption.
    (1) Due to the arrangement of the hepatic portal vein, foodstuffs are picked up (absorbed) from the body and go to the liver before going to systemic circulation.
  ii. **Detoxification:** The liver houses enzymes that break down both toxins from within the body and ingested toxins. In older people the activity of these liver enzymes is lower; thus dosages should be adjusted accordingly.
  iii. **Synthesis of Plasma Proteins:** All of the plasma proteins except for the gamma globulins are synthesized *de novo* in the liver.
  iv. **Storage:** Glycogen, fats, vitamins, minerals (iron, copper) are all stored here.
  v. **Involved in activation of Vitamin D**
  vi. **Removal of bacteria and worn out RBCs. Kupffer cells** are the macrophages that participate in this in the liver (part of the **reticuloendothelial system**).
  vii. **Excretion** of cholesterol and bilirubin
  viii. **Production of bile salts:** Important in fat digestion
  ix. **Alters hormone function:** Somatomedins, etc.

II. Anatomy

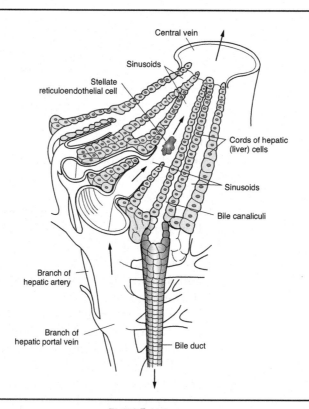

FIGURE 11.2

a. **Lobule:** The basic functional unit of the liver (kind of donut-shaped). We have 500 to 1,000 lobules in the human liver. These are constructed around the central vein (hole in the donut).
   i. The **central vein** leads to the hepatic vein and the vena cava.
   ii. Cells are arranged like spokes on a wheel around the central vein (in the middle).
b. **Microanatomy**
   i. **Circulatory supply from two sources:** Branch of the hepatic portal vein and branch of the hepatic artery (locate both in above picture and notice blood flow direction).
      (1) Both drain into the elongated **sinusoids** (these run between the cords of the cells). "Sinusoids" are just the capillaries of the liver.
   ii. Rows of **hepatic cells of the liver are bathed on either side by the sinusoids.**
   iii. **Bile canaliculi:** These drain bile out of the liver and into the gallbladder.
   iv. **Sinusoids house the Kupffer cells** (mentioned above).
c. **Blood Flow**
   i. **Anatomy:** Blood flows into the liver via the hepatic artery and hepatic portal vein → Sinusoids → Central vein → Drains to hepatic vein → Vena cava → Heart
   ii. **Dynamics & Pressures:** There is very little resistance to flow through the sinusoids. If you increase resistance, edema usually results. Pathologically, the resistance is increased due to **cirrhosis** (the destruction of tissue and subsequent replacement with connective tissue).
      (1) Edema of the liver usually results from cirrhosis because the increased pressure causes more fluid to accumulate in the interstitium around the liver. This edema in the body cavity (ascites) causes a swollen abdomen.
      (2) Due to increased portal resistance, we may also see a "varicocele" upstream. That is to say, the increased sinusoid pressure may cause pooling in the capillaries that lead to the portal system (may see varicose vein in esophagus, etc.).

III. **Liver, Gall Bladder, Duodenum—Relationships**
   a. **Release of Bile:** Hepatocytes continuously make bile. It goes to the bile canaliculi and is drained by the hepatic duct. This leads to the cystic duct and then the gallbladder. The bile is stored within the gall bladder. We secrete up to 1 liter or so a day of bile, but the gallbladder does not hold this much. Thus the gallbladder reabsorbs water and concentrates the bile. The gallbladder contracts and forces the bile into the common bile duct, then into the duodenum. Secretion of bile into the duodenum occurs when food is in the SI. **This release is regulated by two hormones.**
      i. **Cholecystokinin (CCK):** Mucosal cells of the SI produce this in response to the presence of partially digested food (especially fats).
         (1) This causes the gallbladder to contract and the sphincter to relax so that bile can be secreted into the duodenum.
      ii. **Secretin:** Has a small role in bile secretion, although it is not nearly as important.

IV. **Constituents of Bile:** Bile is a digestive secretion *and* a mixture of metabolic waste! Bile includes water, bilirubin, cholesterol (this is used originally to make bile salts in liver), lecithin (a phospholipid), and bile salts.
   a. **Bilirubin:** See RBC section (in Chapter 10). It is a waste product from the heme portion of the RBC.

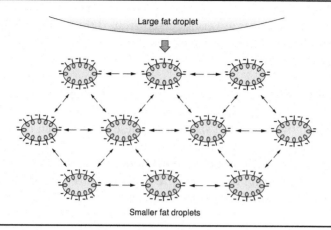

Large fat droplet

Smaller fat droplets

FIGURE 11.3

b. Bile Salts
  i. Function in Fat Digestion: The chemical structure has a water-soluble end and a lipid-soluble end. This allows bile salts to serve two very important digestive functions.
  (1) Emulsification: Fat globules are digested in the intestines by lipases (which are water soluble). Lipases can act only on digesting the surface of a fat globule. So when bile salts break apart a fat globule into smaller balls as shown in the above figure, the surface area on which the lipases can work is greatly increased.
  (2) Micelle formation: When fats are digested by pancreatic lipases they are converted from triacylglycerols (TAG) into free fatty acids (FFA) and monoglycerides (MAG).

  Lipase + TAG → MAG + FFA

  The environment of the lumen of the SI is an aqueous one, and free fatty acids are not water soluble. If fatty acids accumulate (because they are not soluble and thus not readily transported), the reaction slows down or stops (due to the law of mass action—a buildup on the right side of the above reaction). We must transport monoglycerides and the fatty acids to continue digestion. Bile salts form **micelles** (fat "taxi cabs") to transport the fats. Bile salts have a lipid-soluble portion and a water-soluble portion. A micelle is a huddle of these salts together so that the water-soluble moiety points toward the aqueous environment and the lipid-soluble moiety points toward the fat particle it has surrounded (we have effectively made the fat molecule water soluble). We can now transport the fat to the intestinal mucosa, where the monoglycerides and free fatty acids are absorbed and incorporated into the lymph system as TAGs in a package known as chylomicrons. Micelles can then go back and pick up more and continue the cycle.

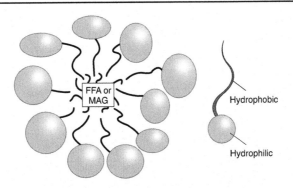

Hydrophobic

Hydrophilic

Formation of a micelle to transport a MAG or FFA

FIGURE 11.4

ii. **Fate of Bile Salts:** A number of bile salts are recycled. Within **the ileum** of the SI, bile salts are reabsorbed and put into the hepatic portal circulation. They eventually make it back to the liver to be used again; 95% of bile salts that are secreted are recycled, while about 5% are lost in the feces (do not be confused, the bile *per se* is not recycled).

c. **Cholesterol**
   i. **Function:** Has no real function in bile (a leftover from making bile salts)
      (1) The precursor in the synthesis of bile salts.
      (2) It is important that cholesterol be maintained at an appropriate ratio/concentration. Cholesterol is lipid soluble (not water soluble), so if we do not have enough bile salts, **large ratios of cholesterol present here can be involved in forming gallstones.**
   ii. **Gallstones:** Cholesterol or bilirubin could precipitate out and form gallstones (in the gallbladder). They may reach a duct (cystic, common bile) and cause a blockage; 10%–20% of the US population will have one at some point. **Treatment:**
      (1) **Sphincterotomy:** If the stone is in a duct, then we may try this procedure. This is done by placing a tube into the SI and up into the duct to try to remove the stone.
      (2) **Lithotripsy:** Use sound waves (ultrasound) to blast the stones. The goal is to fragment the stone so that it may then move along and pass.
      (3) **Pharmacological:** ACTIGALL (ursodiol) helps dissolve stones. This drug is much better at dissolving cholesterol stones than stones that contain bile pigments.
      (4) **Could remove the gallbladder.** Bile will still be produced and will trickle into the SI all of the time, instead of being regulated for fat digestion.

V. Metabolic Functions of the Liver
a. Carbohydrate metabolism

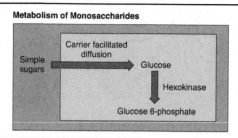

Metabolism of Monosaccharides

FIGURE 11.5

   i. **Metabolism of monosaccharides:** Plasma glucose levels do not change a great deal. Normal plasma glucose is around 70–110 mg/dL. Postprandially, this may climb to >120–150. The hepatic portal vein comes laden with glucose. The hepatocytes take up much of this glucose and convert it to glucose-6-phosphate. This is useful because the uptake into the cell is carrier facilitated (GluT [glucose transporter] 2), and this phosphorylation allows more glucose to be taken into the cell. Also, as we will see, insulin stimulates the activity of hexokinase IV ("glucokinase"), which catalyzes this phosphorylation.

   ii. **Gluconeogenesis:** "The production of new glucose from noncarbohydrate precursors." If you were to decrease your food intake, plasma glucose levels would remain constant because the liver is involved in gluconeogenesis.

   iii. **Glycogen metabolism:** Glycogen is the predominant carbohydrate storage form in the liver. Postprandially we store excess glucose in this way (also may convert some to fat in the liver). When we are starving, we can then break down glycogen to glucose that we can use.

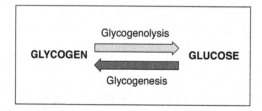

FIGURE 11.6

b. Protein Metabolism
   i. **Plasma Protein Synthesis:** All but the gamma globulins are formed in the liver (gamma globulins are formed in lymphatic tissue).

   ii. **Transamination:** Synthesis and this interconversion of amino acids happens in the liver. We can synthesize all but the nine essential amino acids (these must come from the diet).

   iii. **Urea Synthesis:** Very Important! During the metabolizing of proteins you produce nitrogen and then form ammonia. Ammonia is neurotoxic and must be removed! Thus the liver converts NH_3 to urea in a process called the "ornithine cycle."

c. Lipid Metabolism
   i. **Synthesis of lipoproteins:**

      (1) Types
- a. **Chylomicrons:** TAGs from diet (incorporated in lymph).
- b. **VLDLs:** Very-low-density lipoproteins; made in the liver (TAG rich).
- c. **LDLs:** Low-density lipoproteins. Made in the liver. These transport cholesterol from the liver to various organs. Usually considered the "bad" cholesterol.
- d. **HDLs:** High-density lipoproteins. Made in the liver. These transport cholesterol from organs back to the liver. Usually considered the "good" cholesterol.

      (2) Application in cholesterol biology: Atherosclerosis
- a. LDLs (in general) are depositing cholesterol to blood vessel walls, while HDLs (in general) seem to remove cholesterol from the blood vessel walls and bring it back to the liver.

ii. **Synthesis of cholesterol:** We do need some cholesterol (it is involved in the synthesis of Vitamin D, bile salts, adrenal corticoids, and sex steroids). **We make cholesterol *de novo* in the liver (most of the cholesterol in the body) or absorb it from our diet.**

iii. **Synthesis of phospholipids:** Phospholipids make up the cell membrane structure.

iv. **Hypolipidemic drugs:** Drugs that decrease the plasma cholesterol levels.

      (1) ZETIA: Inhibits the intestinal absorption of cholesterol.

      (2) LIPITOR and ZOCOR: These both inhibit *de novo* cholesterol synthesis in the liver. These are both prescribed *very* often. These drugs are called statins. The main mechanism of action is that by inhibiting cholesterol synthesis in the hepatocytes, the LDL receptors on hepatocytes are upregulated and more LDL is removed from the blood.

      (3) VYTORIN: A combination which contains Zetia and a statin.

VI. Storage by Liver: Vitamin A, D, K (all fat soluble) and some minerals, such as iron

VII. Detoxification by Liver
- a. **Detoxification** of hormones, alcohol, and a number of drugs
- b. **Drugs** are foreign to the body, so it tries to destroy them.
- c. Many enzymes are involved in the detoxification of drugs in the liver. Many things influence these enzymes, such as age (older people and newborns both have impaired enzyme function), cigarette smoking, and other drug interactions. **Drug dosage must account for this!**
- d. **Example:** Coumadin is metabolized by P-450 oxidase. The drug **phenobarbital** (like many barbiturates) induces P-450 enzymes (increases activity). Thus, when adding phenobarbital to a patient's coumadin regimen, the coumadin dosage must be increased as well.

VIII. Liver Modulation of Hormone Action
- a. **Thyroxin:** Thyroxine (T4) is converted to T3 in the liver.
- b. **Growth Hormone:** Liver produces somatomedins, which actually mediate the growth hormone effect (also known as IGFs).

IX. Pathologies
- a. **Hepatitis:** An inflammation of the liver caused by a virus.
  - i. **HAV:** Hepatitis A virus or **infectious hepatitis. The most common type in the U.S.** This is transmitted by a **fecal-oral route** (usually in children without good hygiene) or in shell fish (feces in the water contaminate the fish). This will run its course in a few months.
  - ii. **HBV:** Hepatitis B virus or **serum hepatitis.** This is caused by **intimate contact with the body fluids** (contaminated body fluids). Also may be transmitted by contact with contaminated blood products, transfusion, dirty needles, or sexually transmitted. Involves a longer incubation period and is **more serious.**

iii. **HCV:** Hepatitis C virus or **non A–non B**: most commonly transferred by **dirty needles or blood transfusion; may also be sexually transmitted.**
iv. **HDV:** Hepatitis D virus: **affects only people who have had hepatitis B previously.** Hepatitis B and D significantly damage the liver.
v. **HEV:** Hepatitis E virus: usually from contaminated food and water. This type is the most common worldwide (it is rare in the U.S.).

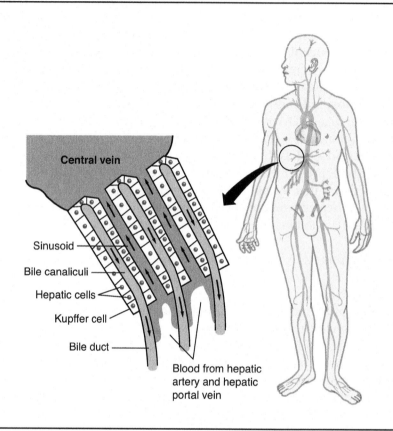

Central vein

Sinusoid
Bile canaliculi
Hepatic cells
Kupffer cell
Bile duct

Blood from hepatic artery and hepatic portal vein

FIGURE 11.7

SUGGESTED READINGS

1. Guyton, A.C., and Hall, J.E. 2011. Chapter 70 in *Textbook of Medical Physiology*, 12th ed. Baltimore: Saunders.

CHAPTER 12

Hemostasis

I. **Vascular Spasm**
   a. If there is a cut or a tear in a vessel, the damaged cells at the site will release several vasoconstrictive agents (serotonin, thromboxane A_2) to reduce blood loss. This is seen more in the arterioles because they have a more prominent muscle wall.
      i. **Why do you get a lot of bleeding when you cut yourself while shaving with a very sharp razor? Is there much tissue damage there?**

II. **Formation of the Platelet Plug**
   a. **Platelets:** Formed elements that are fragments of megakaryocytes
      i. **Platelets normally flow unimpeded in the circulatory system.** As long as they encounter smooth endothelium, they continue to flow smoothly. If damage occurs, they become sticky and stick to one another as well as the damaged site.
      ii. If the inner endothelium of the blood vessel is damaged, the platelets become sticky and **attach to the underlying collagen fibers in the vessel.**
      iii. **The platelets:**
         (1) **Begin to grow processes**
         (2) **Become sticky**
         (3) **Stick to surfaces (collagen)**
         (4) **Stick to each other**
         (5) **Signal to other platelets to "pile on"**
   b. **Signals**
      i. **ADP:** This is released from the damaged site. ADP causes the surfaces of neighboring platelets to become sticky.
      ii. **Thromboxane A_2** promotes aggregation of platelets. This encourages other platelets to become sticky and stick together for a thicker plug.

iii. These occur only where there is damage to the tissue.
iv. Natural endothelium produces **prostacyclins.** These inhibit aggregation of the platelets.
v. Our bodies are constantly making micropunctures in the vasculature, but they are plugged with platelets before any blood can escape.

III. Coagulation

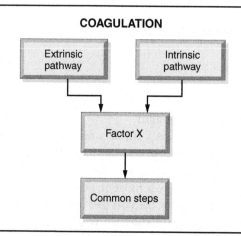

COAGULATION

FIGURE 12.1

a. **Introduction:** Coagulation is a very powerful mechanism. It is mediated at several steps by coagulation factors (most of the factors are made in the liver and are indicated by Roman numerals I–XIII). These coagulation factors usually circulate in the blood in the inactivated state.
   i. There are two different cascade systems: **intrinsic and extrinsic.** These both lead us to the **common pathway,** which begins with **Factor X (prothrombin activator).**
      (1) After Factor X is reached, the steps are common to all.
   ii. **Intrinsic:** All the factors needed **exist in the blood itself.**
      (1) Blood **coagulates intravascularly** or **if in a test tube.**
      (2) Damage → Many steps (Ca^{++} needed in some places!) → Factor X
      (3) There are **more steps in the intrinsic than extrinsic pathway.**
   iii. **Extrinsic:** Involves something from outside the circulatory system (usually in the interstitial fluid). This is how blood clots in extracellular fluid (a bruise). This is a much shorter pathway!
      (1) If blood escapes into the interstitial fluid, then this is responsible.
      (2) Tissue damage → **Tissue thromboplastin** → Factor X
b. **Formation of Prothrombin Activator (Factor X):** This converts prothrombin to thrombin.
   i. Prothrombin is a plasma protein made in the liver.
c. **Conversion of Prothrombin to Thrombin:** In order to convert prothrombin to thrombin, you **need Ca^{++} and Vitamin K.**
   i. The liver is continually synthesizing prothrombin (we have no more than a 1-day supply of this at any time).
   ii. **Vitamin K:** In the 1900s farmers in the northern United States had cows that were hemorrhaging to death. The cows were eating a sweet clover that had a fungus that produced **dicoumarol** (an anti–Vitamin K). This prevented the formation of thrombin.

(1) The cows were not able to produce prothrombin, so they bled to death from the microvasculature punctures that occur everyday.

(2) Dicoumarol, warfarin (COUMADIN) (all anti–Vitamin K agents). **A large dose of this can be used as rat poison** because it induces a fairly slow death (days). This way the rat will pick up the poison and then go die elsewhere; the rats do not associate where they are picking up the poison from with death (they die in the next 2–3 days from microvasculature punctures, not immediately).

   d. Conversion of Fibrinogen to Fibrin: Thrombin is a protease that shortens fibrinogen into fibrin.

     i. Fibrin is a monomer that polymerizes with other fibrin molecules to form a loosely connected meshwork to trap the formed elements.

     ii. Thrombin also activates **Factor XIII (fibrin stabilizing factor).** This strengthens the bonds between the fibrin threads and makes the clot more permanent.

       (1) When the threads are pulled together, they squeeze out **serum.**

   e. Fate of the Clot (Two Fates)

     i. The clot can be dissolved by plasmin: Plasminogen is incorporated into the clot. This can be activated to plasmin, which then dissolves clots.

     ii. Fibroblasts can invade the clot and replace it with new tissue. Sometimes this new tissue leaves a scar because there is more connective tissue with the fibroblasts.

   f. Conclusion

     i. How do you keep the factors localized once the pathway starts?

     ii. Why doesn't the clot spread? Thrombin is dangerous because fibrinogen is everywhere.

       (1) Thrombin is the limiting factor; 90%–95% of thrombin is used up locally.

       (2) The body also produces **anti-thrombin heparin cofactor,** which destroys thrombin. We make heparin from mast cells. This heparin enhances the activity of anti-thrombin heparin cofactor. This anti-thrombin heparin cofactor destroys any thrombin that starts circulating other places.

IV. Anticoagulants and Fibrinolytic Drugs

   a. Anticoagulants: Different drugs that prevent the blood from clotting. These drugs all work to prevent clots, *not* to break up clots that have already formed. We will classify each into two categories.

     i. *In vivo:* Safe to use in a living organism (can use in test tube too)

     ii. *In vitro:* To be used only in a test tube (would kill an organism)

       (1) Heparin (*In vivo*)

         a. It is made **mostly by the mast cells** and blocks the coagulation in several places **(intrinsic and extrinsic).**

         b. Expensive

       (2) Warfarin (COUMADIN) (*In vivo*)

         a. Anti–Vitamin K (thus prevents common pathway)

         b. Used for deep thromboses/to arrest clotting due to surgery

         c. Much cheaper than heparin

       (3) Citrates and Oxylates (*In vitro*)

         a. These tie up the Ca^{++} needed in the intrinsic system.

         b. If a drug ties up ionic Ca^{++} then it functions as an anticoagulant. **These both precipitate Ca^{++}.**

      c. Experiments in rats showed that Ca^{++} "guards" the Na^+ channels (to slow action potentials), so **if you remove the Ca^{++}, then the result is spontaneous action potentials and death from tetany.**
    (4) clopidogrel (PLAVIX)
      a. Binds ADP receptor on platelet to prevent platelet plug formation
  b. Fibrinolytic Drugs: Drugs that actually break up formed clots
    i. Streptokinase (STREPTASE)
    ii. (ACTIVASE): Recombinant–structured "plasminogen activating factor." Thus, it causes plasmin formation and clots get dissolved. Activase is expensive! These drugs may be given to patients as they are having a heart attack or stroke, to dissolve the clot (through an angioplastic catheter).

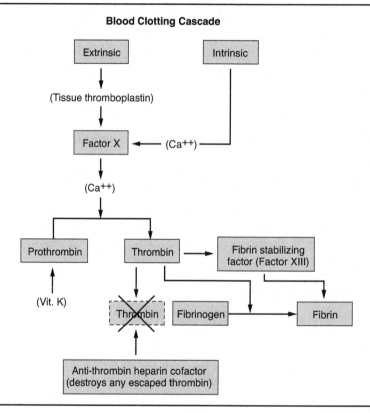

FIGURE 12.2

V. Pathologies
  a. Hemophilia: Lack of one or more of the clotting factors
    i. 80% of the people suffering from this are missing Factor VIII.
  b. Thrombus & Embolus
    i. Thrombus: Intravascular blood clot. This typically happens in the veins of the leg (where blood pools more). If it breaks away, it becomes an **embolus.** An embolus is dangerous because the next capillary system encountered is the pulmonary system—where it may cause a **pulmonary embolism** (lodging in and blocking a pulmonary capillary).

c. **Disseminated Intravascular Clotting**
   i. Clotting of the coagulation system becomes overstimulated by some acute illness (e.g., septic shock), and a large number of clots form. These clots are deposited throughout the capillary circulation.
   ii. **The result is that you run out of clotting factors and bleed to death because you have essentially "used up" your clotting factors.**
d. **Thrombocytopenia Purpura**
   i. There are too few platelets, so platelet plugging is impaired.
   ii. These patients, as a result, have purple blotches/spots (bruises) on their bodies.
   iii. These patients must rely to a great extent on tissue thromboplastin (in the interstitium) to stop the bleeding (the extrinsic pathway).

SUGGESTED READINGS

1. Guyton, A.C., and Hall, J.E. 2011. Chapter 36 in *Textbook of Medical Physiology*, 12th ed. Baltimore: Saunders.

CHAPTER 11 AND 12 QUESTIONS

1. _____ Bile salts
   A. Are used to make micelles
   B. Are used to synthesize lipoproteins
   C. Digest fats
   D. All of these
   E. Two of these

2. _____ Factor X
   A. Is the final step in the intrinsic pathway
   B. Is the final step in the extrinsic pathway
   C. Coverts fibrinogen into fibrin
   D. All of these
   E. Two of these

3. _____ Most common hepatitis in the U.S.
   A. HAV
   B. HBV
   C. HEV
   D. DVD
   E. MTV

4. _____ A function of the liver
   A. Glycogenesis
   B. Glycogenolysis
   C. Gluconeogenesis
   D. All of these
   E. Two of these

5. _____ Would likely be found in the hepatic portal vein
   A. Bile salts
   B. Bile
   C. Chylomicrons
   D. All of these
   E. Two of these

For questions 6–8, choose one of the following:
   A. Increases or is greater than
   B. Decreases or is less than
   C. Has no effect or is equal to

6. _____ The effect of prostacyclins on the likelihood of platelets to aggregate

7. _____ The number of steps in the extrinsic mechanism of blood coagulation, as compared to the number of steps in the intrinsic mechanism

8. _____ The effect of Coumadin on the synthesis of prothrombin activator

CHAPTER 13

Endocrinology I: Introduction, Hypothalamus, Pituitary

I. **Intercellular Communications:** Different sorts of molecules have different chemical messengers they send between cells to communicate.
    a. **Paracrines:** These affect an adjacent cell (do not go very far).
        i. They do not get into the circulatory system.
        ii. This is chemical diffusion to a neighboring cell.
        iii. Short-lived
    b. **Neurotransmitters:** Paracrine-type secretion (presynaptic and postsynaptic concept)
        i. These are synthesized in the neuron and released to postsynaptic structure.
        ii. Really, these are paracrines, but they are specialized.
    c. **Hormones:** These are usually produced by the endocrine cell in response to a stimulus.
        i. Carried to the target tissue (usually specific) by the circulatory system
        ii. Target has receptors on it that recognize the hormones.
        iii. These are highly potent.
    d. **Neurohormones:** Borderline molecules such as norepinephrine and epinephrine from the adrenal medulla.
    e. **Autocrine Secretions:** This is the most unusual. A single cell releases a substance that then has an effect on the original cell that released it.
    f. **Pheromone:** A "chemical" secreted into the external environment to act on a member of the same species.

Nervous System	Endocrine System
Hard wired (electrically)	Chemical communications
Communicate short distances	Communicate long distances
Faster transmission (milliseconds)	Slower transmission (minutes → days)
Shorter-lived response	Longer-lived response
Action potentials generated	

II. **Feedback Mechanisms**
   a. **Simple:** These normally involve a negative feedback system that is nonhormonal.
      i. **Nonhormonal feedback:** Endocrine cells release the hormone to the target tissue. This causes a biological response that serves to turn off the secretion of the hormone (the hormone itself does *not* cause the turn-off).
      ii. **Example: Glucose and insulin regulation**—Increased plasma glucose causes insulin to be secreted. This causes plasma glucose levels to decrease, and then insulin secretions also decrease.

   b. **Complex (multilevel):** These systems also involve negative feedback, but the hormone itself causes the stimulus to shut off the secretion, **not the biological response**.

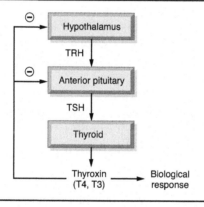

FIGURE 13.1

   i. **Example:** The hypothalamus releases thyrotropin releasing hormone (TRH), which stimulates the anterior pituitary gland to release thyroid stimulating hormone (TSH), which then goes to the thyroid gland to cause release of T4 and T3, which results in a biological response.
   ii. **T3 and T4 will come back to inhibit** the hypothalamus and the anterior pituitary to shut off the system (free T4 is most strongly coupled with TSH levels).

III. **Chemical Nature of Hormones**: Three main chemical classes
   a. **Amino Acid Derivatives:** These are formed by the simple conversion of common amino acids (e.g., thyroxine). These are typically very small, fairly unusual, and hydrophilic.
   b. **Peptide and Protein Hormones:** These are larger peptides and larger protein hormones. **Most hormones fall into this category. These are hydrophilic.**

c. **Steroids:** These are lipid soluble, hydrophobic, and lipophilic. These are all synthesized from cholesterol.
   i. All hormones are transported by the blood; if something is hydrophobic then **you have to attach a plasma protein to it to transport it.**

IV. **Synthesis**
   a. **Peptides and Proteins:** These are made in the **cell ribosomes** (the **pro-hormones**) and then sent to the **Golgi apparatus** (which modifies it and makes it the real thing). We do have **some storage of these hormones.**
      i. When it is appropriate (the cell is stimulated), the cell will exocytose the hormone into the plasma.
   b. **Steroid:** These are made from cholesterol (transported via LDLs to cell). When the steroid is made in the cell, it is released (**virtually no storage**). Because of this, **it takes longer to get a steroid hormone on site than it does a peptide hormone.**
      i. Steroids are made and released on the appropriate signal.
      ii. Steroids pass right through the membrane.
      iii. Steroids usually take their action on the effector cell in the nucleus.
      iv. Steroids also travel with a transport protein so that they do not get filtered and are more soluble.

V. **Mechanism of Action**
   a. **Proteins**
      i. These utilize some 2nd messenger: These cannot get through cell membranes easily, so their response is mediated via a receptor.
      ii. Some proteins **change the permeability of the membrane.**
   b. **Steroids**
      i. These pass right through the membrane.
      ii. The receptor is usually in the nucleus.
      iii. These bind receptors and alter the DNA machinery of the nucleus.

(The feedback loops of hormones are not necessarily concrete; some things supersede and intervene in the above models. Some things complicate these simple systems, like a single endocrine gland produces more than one hormone, or a single hormone is produced by more than one gland, or a target is influenced by more than one hormone, etc.)

HYPOTHALAMUS—PITUITARY RELATIONSHIP

I. **Introduction:** These two structures form the most complex and the most dominant portion of the endocrine system. The entire structure weighs only 500 mg (hypothalamus) and 10 mg (pituitary gland). The pituitary gland (or the **hypophysis**) is located at the base of the brain and is protected by the sphenoid bone (and a saddle-like depression on the sphenoid bone called the sella turcica). Historically, every time doctors/scientists tried to remove it, the animal died. This is not because it was vital but because it is so well protected and hard to get to (it was the surgery that killed the animals, not the loss of the pituitary). The hypophysis is closely associated with the hypothalamus. The pituitary gland is divided into two separate portions (two parts of same gland), which are very different and have different embryological origins.

a. **Adenohypophysis:** Synonym for anterior pituitary
b. **Neurohypophysis:** Synonym for posterior pituitary; connected to the hypothalamus

II. **Embryology**
    a. **Anterior Pituitary:** Derived from **Rathke's Pouch** (from the **developing mouth** in an embryo)
    b. **Posterior Pituitary:** From the **developing brain** of the embryo (a projection called the **infundibulum gives rise to the posterior pituitary**)
       i. Rathke's pouch **breaks off and migrates up** to the infundibulum and becomes a part of it.

III. **Adult Anatomy:** Rathke's pouch surrounds the infundibulum until adulthood.

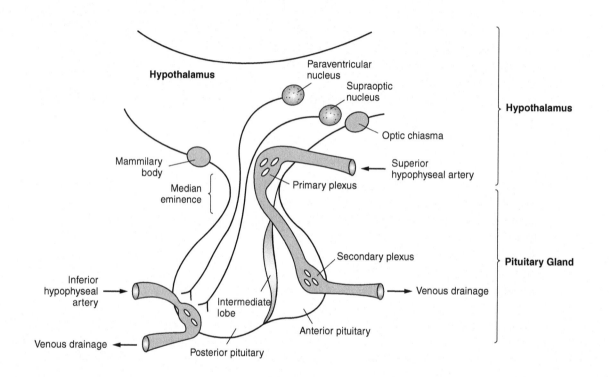

FIGURE 13.2

    a. **General Anatomy** (see above figure) and notes
       i. The intermediate lobe is not well developed in humans.
       ii. Pars tuberalis is part of the anterior pituitary that is wrapped around the infundibular stalk.
       iii. The **supraoptic nucleus** (ADH) and **paraventricular nucleus** (oxytocin) both communicate neurally with the posterior pituitary.
       iv. The hypothalamus communicates with the anterior pituitary via another portal system.
    b. **Anterior Pituitary Cell Types and Hormones Secreted:** Within the anterior pituitary we find several different cell types (that each make different hormones).
       i. **Somatotrophs (40%–50% of cells):**
          - Make human growth hormone (**hGH**)
          - Are the most prevalent of cells

ii. **Lactotrophs (10%–15% of cells):**
  - Make **Prolactin**
iii. **Corticotrophs (15%–20% of cells)**
  - Make adrenocorticotrophic hormone (**ACTH**)
  - Make melanocyte-stimulating hormone (**MSH**)
  (MSH is important in some species, **but not in humans**—these two hormones are very similar to each other. Humans exposed to high levels of MSH have a response as if they were exposed to ACTH.)
iv. **Thyrotrophs (3%–5% of cells)**
  - Make thyroid-stimulating hormone (**TSH**).
v. **Gonadotrophs (10%–15% of cells)**
  - Make follicle-stimulating hormone (**FSH**) and luteinizing hormone (**LH**).

IV. **Anatomical and Physiological Relationships between the Hypothalamus and Pituitary**
  a. **Neural relationship and posterior pituitary**

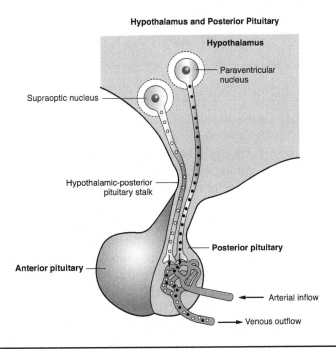

FIGURE 13.3

i. The infundibular stalk contains bundles of nerve fibers, which originate in the hypothalamus. These end in the posterior pituitary next to the capillary bed.
  (1) **From the supraoptic nucleus:** Produces predominantly ADH
  (2) **From the paraventricular nucleus:** Produces predominantly oxytocin. These are both made in the respective nucleus of the hypothalamus and are transmitted through the neuron for storage in the posterior pituitary and eventual release.

b. Circulatory relationship and anterior pituitary

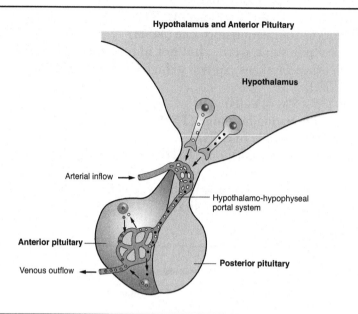

Hypothalamus and Anterior Pituitary

Hypothalamus

Arterial inflow

Hypothalamo-hypophyseal portal system

Anterior pituitary

Venous outflow

Posterior pituitary

FIGURE 13.4

i. **This is a circulatory connection (not neuronal).**
   (1) Thus, the anterior pituitary makes its own hormones (these cells are *not* innervated by cells from the hypothalamus). **The hypothalamus releases inhibiting or releasing factors (hormones) into a portal system that travels to the anterior pituitary and causes it to secrete.** Hormones from the hypothalamus are secreted into the primary plexus. They then travel down the hypophyseal portal vein to the secondary plexus to be secreted into the anterior pituitary to affect the target cell type (and thus cause the release of the proper hormone into systemic circulation).
ii. **Releasing/Inhibiting Hormones from the Hypothalamus**
   (1) Thyrotropin-releasing hormone (**TRH**): Stimulates the release of **TSH** and **Prolactin**
   (2) Corticotropin-releasing hormone (**CRH**): Stimulates the release of **ACTH** and **MSH**
   (3) Gonadotropin-releasing hormone (**GnRH**): Stimulates the release of **FSH** and **LH**
   (4) Growth hormone–releasing hormone (**GHRH**): Stimulates the release of **hGH**
   (5) Growth hormone–inhibiting hormone (**GHIH** or **somatostatin**): Inhibits the release of **hGH**
   (6) Prolactin-inhibiting hormone (**PIH**): Inhibits the release of **prolactin** (this is **probably dopamine**).

Hypothalamus–Pituitary Review

Hormone from Hypothalamus	Affected cell type in Anterior Pituitary	Hormone from Anterior Pituitary
TRH	Thyrotrophs	TSH Prolactin
CRH	Corticotrophs	ACTH MSH
GnRH	Gonadotrophs	FSH LH
GHRH	Somatotrophs	hGH
GHIH	Somatotrophs	Inhibits hGH release!
PIH	Lactotrophs	Inhibits prolactin release!

V. **Anterior Pituitary Hormone Functional Overview (these are proteins)**
   a. **FSH: Follicle stimulating hormone**
      i. Causes growth and maturation of the ovarian follicles
      ii. Influences estrogen secretion from the follicle
      iii. Stimulates sperm production
   b. **LH: Luteinizing hormone**
      i. A surge of LH is needed for ovulation.
      ii. Causes formation of the **corpus luteum** (follicle reorganizes after ovulation to form the corpus luteum)
      iii. Causes secretion of testosterone by the Leydig cells
   c. **ACTH: Adrenocorticotropic hormone**
      i. Stimulates the production of the glucocorticoids by the adrenal cortex (primarily cortisol in humans)
   d. **TSH: Thyroid stimulating hormone**
      i. Causes growth/maintenance of the thyroid tissue
      ii. Stimulates the secretion of the thyroid hormones
   e. **Prolactin**
      i. Causes secretion of milk in the mammary glands (males do produce prolactin, but with no known function)
      ii. Secretion: There are alveolar sacs in the breast that are lined with secretory epithelial cells (synthesize and secrete milk into the alveoli). Prolactin causes the secretion of the milk into the sacs, *not* the "letdown" of the milk to the nipple.

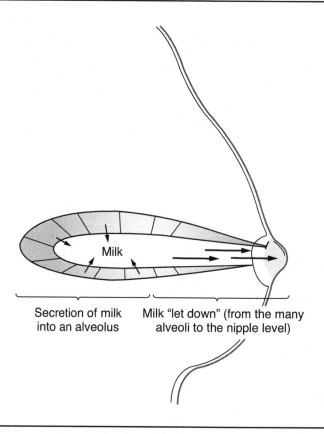

Milk

Secretion of milk
into an alveolus

Milk "let down" (from the many
alveoli to the nipple level)

FIGURE 13.5

f. **hGH: Human growth hormone**
  i. Role in body growth and the regulation of metabolism (see hGH section)
g. **Melanocyte stimulating hormone (MSH)**
  i. Associated with the **intermediate lobe** (in adults). This *is* part of the anterior pituitary.
  ii. **No significant role in humans**, although it may play a role in appetite suppression.
  iii. MSH has a role in **camouflaging some animals.** MSH favors the distribution of melanin throughout melanocytes. If melanin is distributed throughout the cell, the cell takes on a darker color.
  iv. **ACTH influence:** ACTH is MSH's "close cousin" (MSH is part of ACTH—from the POMC gene). Patients who have a pathologically high ACTH may begin to take on a darker color (due to similarity to MSH).

VI. **Posterior Pituitary Functional Overview (these are also peptides)**
  a. **ADH** (arginine vasopressin in humans)
    i. Produced by the **supraoptic nucleus in the hypothalamus** and transmitted down neurons to be stored and released from the posterior pituitary. ADH has a role in urine production and regulating BP (see Chapter 8, Water Regulation and Kidney Function).
  b. **Oxytocin**
    i. Produced by the **paraventricular nucleus in the hypothalamus** and transmitted down neurons to be stored and released from the posterior pituitary. This stimulates milk "let-down" and may also play a role in parturition.

ii. **Milk letdown:** Milk is secreted into the alveolar sac, but it then has to go through a ductile system to the nipple so that it may be suckled. The cells located right under the sac are myoepithelial cells (smooth muscle) that have oxytocin receptors on them. When stimulated, these receptors cause contractions of the myoepithelial cells so that the milk is let down the lactiferous ducts to the nipple. Oxytocin is released in response to sucking by the infant at the nipple (a **neuroendocrine reflex:** neuro at the afferent side and endocrine at the efferent side). This reflex can be easily conditioned and other stimuli may come to cause milk let-down. The unconditioned stimulus would be the suckling from the infant, but the conditioned stimulus could be hearing the baby cry.
   iii. **Oxytocin has been implicated in parturition:** There are oxytocin receptors on the myometrium of the uterus. Oxytocin will hasten parturition in a properly prepared uterus. **Pitocin** is synthetic oxytocin that is used to induce labor. Does oxytocin normally trigger parturition by itself? Probably not:
   (1) Oxytocin levels are elevated only after parturition has already begun.
   (2) Rats with no hypothalamus (and thus with no hypothalamic oxytocin) still parturate on time (it does take them longer to perform that act, though).
   iv. **Oxytocin has also been implicated in assisting sperm transport.** Sperm go from the vagina to the oviducts in less time than just swimming could take (just a few minutes). Sexual intercourse causes oxytocin release in females and is thought to increase the motility of the uterus for sperm travel.

VII. **Pineal Gland:** Not a part of the pituitary gland (is posterior to the thalamus)
   a. At one time it was thought to have had no function in humans, but it does **secrete melatonin** (hormone). **More melatonin is secreted at night** (we appear to have a biological clock that is perhaps mediated by this).
   b. The third eye (or pineal eye) is present in some animals. In humans, the pineal gland is **located close to where the optic tracts cross** (optic chiasm). Implicated in day/night cycles.
   c. Melatonin is **involved in certain functions of cyclic activity** (like the biological clock). Most people are diurnal by nature. Melatonin may be involved in other cyclic processes: affective disorders (seasonal depression), premenstrual syndrome (PMS), puberty onset (how does the body know when to become sexually mature?), and seasonal breeding (animals).
   d. **Melatonin also inhibits FSH and LH.**

SUGGESTED READINGS

1. Guyton, A.C., and Hall, J.E. 2011. Chapters 74 and 75 in *Textbook of Medical Physiology*, 12th ed. Baltimore: Saunders.
2. Koeppen, B.M. and Stanton, B.A. 2008. Chapters 37 and 40 in *Berne & Levy Physiology*, 6th ed. St. Louis: Mosby.

CHAPTER 14

Endocrinology II: Growth Hormone

I. **Introduction**
   a. Human growth hormone (hGH) is needed for humans to grow normally.
      i. It is the most abundant hormone in the anterior pituitary (produced by the somatotrophs).
   b. hGH has its effects at a number of different locations (no one target tissue), such as in the fat tissue (decreased adiposity), muscle tissue (increased lean body mass), and cartilage (increased linear growth via chondrocytes).
   c. hGH also increases the size of a number of tissues (kidney, pancreas, bone, heart, lungs). (Figure 14.3)
   d. hGH is a protein of 101 amino acids. It is called "human GH" because it has lots of species specificity (unlike insulin).
   e. We can now synthesize this with recombinant DNA techniques.

II. **Metabolic Effects**
   a. **Protein Metabolism**
      i. **Growth hormone enhances protein synthesis** by increasing amino acid uptake into the cells, increasing RNA synthesis, and increasing ribosomal activity. The overall effect is an increase in the lean body mass and an increase in the size of other organs.
   b. **Lipid Metabolism**
      i. **Causes lipolysis:** hGH increases lipid breakdown, which leads to an increase in plasma fatty acids.
      ii. **Beta oxidation:** hGH enhances beta oxidation (fatty acid usage as fuel), which leads to an increase in ketone body production (and thus more keto acids in the body that may actually cause metabolic acidosis).
   c. **Carbohydrate Metabolism**
      i. **hGH decreases cellular uptake:** Inhibits the uptake of glucose into the cells, specifically in the muscle and fat cells

    ii. Diabetic effect: hGH increases plasma glucose concentrations, which then causes an increase in insulin secretion from the pancreas and may lead to diabetes.

    iii. hGH also **makes fat and muscle cells more resistant to insulin.**

    iv. hGH is often considered "counter-regulatory" to insulin because of its opposing effects.

III. Growth Hormone and Growth
    a. **Factors affecting growth:** Many things besides hGH affect growth.
       i. **Genetic** background
       ii. **Diet:** Inadequate diet (low protein) can cause improper growth.
       iii. **Freedom from chronic disease and stress:** Chronic illness interferes with growth.
       iv. **Hormones:** Insulin, sex hormones, thyroid hormones
          (1) Growth hormone **does not affect fetal growth** and **does not affect neonates until a few months after birth.**
    b. **Soft Tissue and Bone:** hGH stimulates growth in both (soft tissue includes such organs as kidney, heart, skin, lungs).
       i. **Hypertrophy:** An increase in the size of cells
       ii. **Hyperplasia:** An increase in the number of cells
    c. **Ossification (bone growth): Two types**
       i. **Intramembranous Ossification**
          (1) Occurs in the **periosteum** (membrane on outside of bone).
          (2) hGH **stimulates the osteoblasts** within the periosteum.
          (3) This causes these **bones to grow in diameter.**
          (4) This is **how bones are repaired** when injured.
          (5) The **smaller, flat bones** of the body **solely grow this way.**
          (6) **Example:** In long bones the membrane on the outside is the periosteum. In order for a child to grow, the osteoblasts in the periosteum must lay down bone.
       ii. **Endochondral Ossification (Cartilage):** This occurs mostly in the long bones. First, there is a cartilage precursor or **prototype** (in the fetus you have a cartilaginous prototype) that appears at two centers: the **primary ossification center** and the **secondary ossification center.**

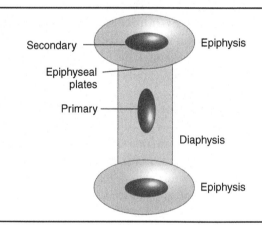

FIGURE 14.1

          (1) At the two centers the bone starts replacing cartilage.
          (2) Neonates already have ossification taking place. In the areas in the bone where there is cartilage, ossification takes place and bone replaces the cartilage.

(3) hGH stimulates the activity of osteoblasts in the periosteum. hGH also **stimulates the prochondrocytes** (cells laying down cartilage).

(4) This process continues until adolescence, when osteoblasts finally overtake the cartilage completely and the epiphyseal plates become fused. hGH can no longer have an effect at these fused sites.

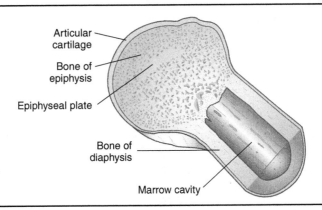

Articular cartilage

Bone of epiphysis

Epiphyseal plate

Bone of diaphysis

Marrow cavity

FIGURE 14.2

c. **Role of Somatomedins (IGFs):** hGH does not always actually bind to the receptors on the target cells. hGH increases the production of somatomedins (also called insulin-like growth factors, IGFs), which in turn act on the target tissues.

   i. **Two types have been identified:** IGF I and IGF II

     (1) Manufactured in the periphery (mainly in the liver)

     (2) We seem to see an increase in the levels of IGF I at puberty, but there is no change in the levels of IGF II at puberty (not sure why).

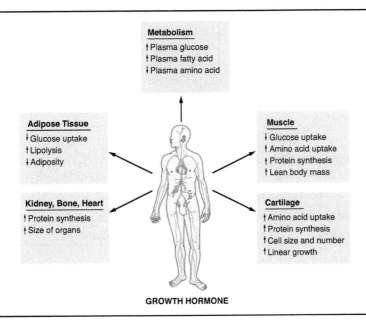

Metabolism
↑ Plasma glucose
↑ Plasma fatty acid
↓ Plasma amino acid

Adipose Tissue
↓ Glucose uptake
↑ Lipolysis
↓ Adiposity

Muscle
↓ Glucose uptake
↑ Amino acid uptake
↑ Protein synthesis
↑ Lean body mass

Kidney, Bone, Heart
↑ Protein synthesis
↑ Size of organs

Cartilage
↑ Amino acid uptake
↑ Protein synthesis
↑ Cell size and number
↑ Linear growth

GROWTH HORMONE

FIGURE 14.3

IV. Regulation: There are more than 20 different things that regulate growth hormone secretion: We do not really know what exactly causes its secretion.

    a. GHRH & GHIH (Somatostatin): Hypothalamus influences hGH secretion. hGH has a releasing hormone and an inhibiting hormone (somatostatin).

        i. These are released into the anterior pituitary via hypophyseal portal system.

        ii. There is periodic production of growth hormone. In adults, hGH is secreted primarily during deep sleep. In growing children, hGH is secreted in bursts during the sleep/wake cycle.

    b. Other factors that stimulate hGH secretion:

        i. Low plasma glucose level

        ii. Low plasma fatty acid level

        iii. High plasma amino acid level

        iv. Sleep

        v. Stress

        vi. Exercise

        vii. Starvation

        viii. Ghrelin

    c. Feedback

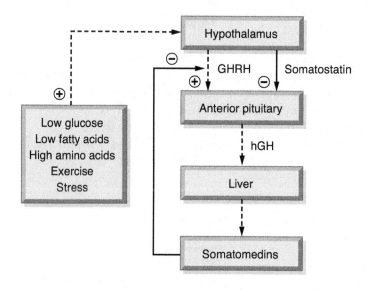

FIGURE 14.4

V. hGH and Insulin "Counter Regulation"

Consumption	Effect on hGH secretion	Effect on somatomedin secretion	Effect on insulin secretion	Bottom line
↑ **Protein intake**	↑	↑	↑	*Not* counter-regulatory
↑ **CHO intake**	↓	NE	↑	Counter-regulatory
Fasting	↑	↑	↓	Counter-regulatory

VI.   **Pathologies**
  a.   **Dwarfism:** There are many different types.
      i.   **Pituitary Dwarf:** A person whose short stature typically is due to too little hGH or GHRH (most common type).
      ii. **Laron Dwarf:** A short-statured person with:
          (1) Normal to high levels of hGH
          (2) An inability to synthesize IGF I
          (3) Mutated hGH receptors
          (4) No increase in IGFs in response to GH.
      iii. **Pygmies:** A group of short-statured people with:
          (1) Low IGF I levels
          (2) Normal hGH receptors
          (3) A deficient number of GH receptors
          (4) Normal IGF II response
  b.   **Poodles (not a pathology, but interesting):** These are inbred to yield three varieties.
      i.   **Growth hormone levels are similar in all.**
      ii. IGF I is highest in **standard poodles** (normal size).
      iii. IGF I is intermediate in **miniature poodles** (small size).
      iv. IGF I is lowest in **toy poodles** (very small size).
  c.   **Gigantism**
      i.   **Pituitary gigantism:** This is associated with too much hGH, generally due to a pituitary tumor. The tumor occurs in children, who thus may grow to be very tall. These patients are also at risk for diabetes mellitus.
  d.   **Acromegaly:** This is associated with too much hGH **after the fusion of epiphyseal plates** (big brow, chin, hands, and feet).

VII. **Future Challenges**
  a.   Now that synthetic hGH is available, **we have medical and ethical questions** that arise as to who should be treated with it.
      - **Burn patients? (hGH helps skin growth)**
      - **Morbidly obese patients? (hGH is strongly lipolytic)**
      - **Elderly? (hGH is anabolic for when tissues begin to waste away)**
      - **Athletes? (hGH is very anabolic and lipolytic)**

SUGGESTED READINGS

1.   Guyton, A.C., and Hall, J.E. 2011. Chapter 75 in *Textbook of Medical Physiology*, 12th ed. Baltimore: Saunders.

CHAPTER 13 AND 14 QUESTIONS

1. _____ The portion of the pituitary derived from Rathke's pouch secretes
   a. Oxytocin
   b. Prolactin
   c. Melanin
   d. All of these
   e. Two of these

2. _____ Somatotrophs produce
   a. GHIH
   b. Somatostatin
   c. Somatomedins
   d. None of these
   e. Two of these

3. _____ Prolactin
   a. Causes the contraction of smooth muscle
   b. Is regulated by an inhibiting hormone
   c. Is made in the posterior pituitary
   d. All of these
   e. Two of these

For questions 4–10, choose one of the following:
   A. Increases or is greater than
   B. Decreases or is less than
   C. Has no effect or is equal to

4. _____ The effect of stimulating the hypothalamic osmoreceptors, on the amount of ADH released into the hypophyseal portal vein

5. _____ The effect of growth hormone on fat deposits

6. _____ The effect of growth hormone on the growth of the kidney after puberty

7. _____ The effect of protein intake on growth hormone secretion

8. _____ The effect of melatonin on FSH secretion

9. _____ IGF I response to growth hormone in a Laron Dwarf, as compared to the response in a normal person

10. _____ The effect of prolactin on milk letdown

CHAPTER 15

Endocrinology III: Reproduction

I. **General Biochemistry**
   a. **Steroid-Producing Cells**
      i. **Adrenocortical:** The adrenal cortex is the site for production of these steroids. There are three zones to the cortex:
        (1) Zona glomerulosa
        (2) Zona fasiculata
        (3) Zona reticularis
        (4) These produce hormones: mainly aldosterone (mineral corticoid) and cortisol (glucocorticoids).
      ii. **Ovarian:** Ovaries produces estrogen and progesterone.
        (1) Follicular stage: This is the first 14 days of cycle; predominantly estrogens are produced.
        (2) Luteal stage: This is the last 14 days of female cycle; predominantly progesterone is produced (and higher than normal levels of estrogen).
      iii. **Testicular:** The Leydig cells of the testes produce androgens in the male.
      iv. **Placental:** The placenta produces steroids as well.

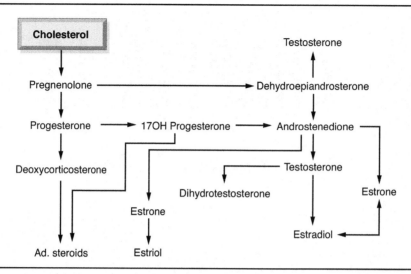

FIGURE 15.1

b. **Pathways**
   i. **Cholesterol is the parent molecule!** It is modified in one way or another to form the different steroids.
   ii. There is very little that is different about each steroid when looking at them, but **biologically they are very different.**

Male Endocrinology

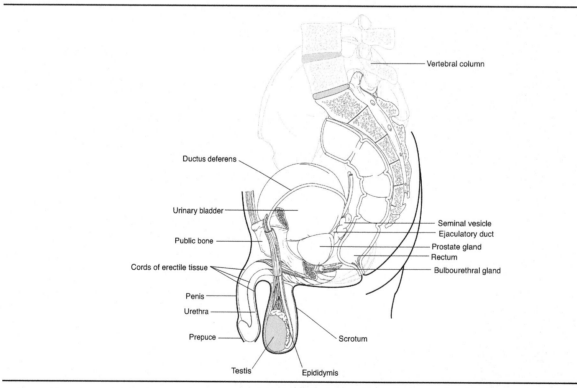

FIGURE 15.2

II. **The Testes:** The primary reproductive structure in the male. The testes have dual functions: **produce sperm (spermatogenesis) and produce androgens (steroidogenesis).**
   a. **Embryology**
      i. The testes develop in the rear of the abdominal cavity. Around the **7ᵗʰ month of gestation,** they begin to migrate and then eventually descend from the abdominal cavity through the abdominal wall. These travel through the **inguinal canal** into the scrotum. The **spermatic cord** is attached to each one.
      ii. Following the descent, the opening of the inguinal canal **closes tightly around the spermatic cord** (which contains blood vessels and nerves).
         (1) If this opening is not closed tightly, an **inguinal hernia** could result. This may be symptomatic or asymptomatic. Part of the gut can bulge through the hole and become strangled, and thus **surgical repair is usually warranted.**
      iii. **The testes must descend into the scrotum** because the temperature in the scrotum is 2–3 degrees lower than the temperature in the abdominal cavity.
         (1) **Spermatogenesis cannot occur at 37°C.**
         (2) **Steroidogenesis can occur at 37°C.**
      iv. There are **two systems** responsible for maintaining the temperature of the testes.
         (1) **Pampiniform Plexus:** This is a countercurrent blood flow to and from the testes. The blood leaving the testes cools the blood coming to the testes.
            a. **Vericocele:** This is like a varicose vein, but in the pampiniform plexus. This disrupts the countercurrent blood flow. As a result, the temperature is too high and that testis is usually sterile (can be fixed).
         (2) **Cremaster muscle:** This muscle can elevate or lower the testes to alter the temperature of them (closer to or farther away from abdominal wall). When the temperature is too high, the muscle relaxes so that the testes will be farther down in the scrotum. When the temperature is too low, the muscle contracts and pulls the testes closer to the body.
            a. **Cryptorchidism:** This pathology is fairly uncommon in full-term pregnancies. It usually occurs in premature babies. With cryptorchidism, the testes do not descend properly during fetal life **(cryptos = "hard to see"; orchios = "testes").**
   b. **Histological Morphology:**
      i. **Longitudinal Section:** There are 200 to 300 cone-shaped lobules within each testis.
         (1) **Seminiferous tubules:** There are one to three of these within each lobule. These are highly convoluted (800 feet in testes; 15,000 feet in bull testes).

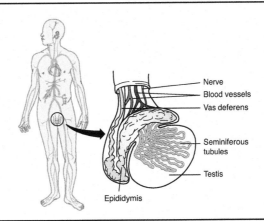

FIGURE 15.3

ii. **Two main cell types within each testis:**
(1) **Seminiferous tubules:** The seminiferous tubules are the sperm-producing apparatus.
   a. **Germ Cells:** These are all in different stages of differentiation or maturation (meiosis).
   b. **Sertoli Cells:** These are in the seminiferous tubules. They are not gametes. **These cells have a number of different functions.**
      i. **Provide the blood-testes barrier:** Immune system learns early in life what is self and what is not self. In the male, spermatogenesis does not occur until puberty. These sperm cells are suddenly new to the body, where the immune system could attack them. Instead, the sertoli cells create this blood-testes barrier.
      ii. **Nourish sperm**
      iii. **Phagocytize germ cells** that do not quite make it (clean up debris).
      iv. **Produce a tubular fluid that helps transport the sperm** (from testes to epididymis).
      v. **Produce "androgen binding proteins."** These proteins allow the testes to maintain an increased concentration of androgens (spermatogenesis has to have a high amount of testosterone in order to occur).
      vi. **Produce the hormone *inhibin.***
      vii. **Increase the activity of the enzyme P_{450} aromatase.**
      viii. **Produce plasminogen activating factor.** This causes the formation of plasmin, which helps "digest the sperm away" from the edge of the lumen of the seminiferous tubules.
      ix. **Increase the synthesis of transferrin** (high concentrations of iron are needed for normal sperm production).
      x. **Produce Mullerian-inhibiting factor (embryonic).**

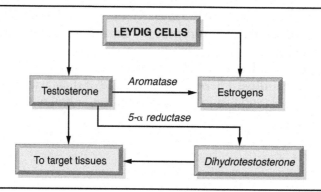

FIGURE 15.4

(2) **Interstitial Cells of Leydig:** These are located in the spaces between the seminiferous tubules. **These cells produce different androgens.**
   a. **Testosterone:** The predominant androgen is testosterone. This can go to the target tissue and affect that tissue, or it can be converted by 5-alpha-reductase into dihydrotestosterone (DHT).
   b. **DHT:** This can also affect target tissues (more potent than testosterone; affects different tissues).
   c. **Estrogens:** Testosterone can be converted into estrogen via P_{450} aromatase. Men have significant quantities of circulating estrogen. Normally, there is no expression

of the estrogens in males because their high levels of androgens mask it. However, when a male begins to take a synthetic testosterone, production of natural testosterone lessens, and he develops a testosterone insensitivity. In this event, the estrogens begin to get expressed. Also, with increased levels of synthetic testosterone, some are more easily converted to estrogens.

Typically these estrogens are expressed as secondary sex characteristics: **Gynecomastia**—rudimentary breast development in a male (in this case due to the above situation).

III.   **The Ductile System:** The **convoluted pathway** takes sperm from the testes to outside the urethra, the **glands** contribute to the semen, and **the penis** is designed to put sperm into the vagina.
   a.   **Epididymis** ("on the twins"): These are set on each of the testes (comma shaped). If you straightened them out, one would be 4–5 meters long.
       i.   The sperm are **stored and mature in the epididymis.**
       ii. (Sperm taken from the seminiferous tubules are infertile.)
   b.   **Vas Deferens:** Leads away from the epididymis.
       i.   **Vasectomy:** This is a surgical procedure of tying the vas deferens on either side; male can still ejaculate, but with no sperm. Requires surgery to reverse (fairly permanent).
   c.   **Seminal Vesicle:** Contributes ~60% of the semen volume (sperm is only ~10% of volume). This secretes a nutrient-rich secretion:
       i.   **Nutrients: Fructose, prostaglandins, and fibrinogen**
   d.   **Prostate:** Contributes ~30% of semen. Located where the urethra comes out of the bladder (surrounds urethra).
       i.   In older men, **prostatic hypertrophy** can cause the urethra to become strangled and/or collapsed. These men have difficulty urinating.
       ii. The prostate actually produces a **thin, milky, alkaline secretion:**
           (1) **The vagina is an acidic environment** so that microbial growth will be retarded. This environment is no good for the motility and life of the sperm.
           (2) Since the **semen is alkaline overall (~7.2),** sperm can survive better in the vagina.

IV.   **Male Sex Act:** Tactile stimulation of the glans penis and/or psychological stimulation.
   a.   **Erection:** A **parasympathetic phenomenon** that causes the dilation of the arterioles that supply the penis. These are the only arterioles that are parasympathetically innervated.
       i.   **Corpus cavernosum:** Spongy tissue within the penis. When this is engorged with blood, it causes an erection.
   b.   **Ejaculation:** A **sympathetic phenomenon.** When the sexual stimuli are sufficient, it causes the contraction of the epididymis, the vas deferens, the prostate, and other glands to cause ejaculation.
   c.   Think: **P**oint (**P**arasympathetic) and **S**hoot (**S**ympathetic).

V.   **Female Sex Act:**
   a.   **Parasympathetic:** Vasodilation of arterioles to clitoris and labia in response to tactile and/or psychological stimulation. Fluid is forced out of vaginal capillaries into the vaginal lumen (lubrication for sex).
   b.   **Sympathetic:** When stimuli are sufficient, the female version of male ejaculation takes the form of contractions of the pelvic musculature.

VI. Testosterone

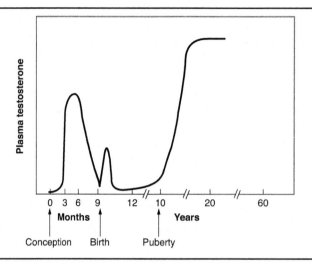

FIGURE 15.5

a. **Changes in levels**
   i. Plasma testosterone levels change as a function of age, as shown in above figure.
   ii. There is a **surge of androgens** during fetal life (in fetal testes) from **8 weeks until 18 weeks.** This is critical for normal development of the male reproductive structures.
   iii. Testosterone levels are low in the neonate after birth.
   iv. Around **2 to 3 months** after parturition, there is another surge in testosterone (the exact role of this surge is unknown).
   v. Over the next 10 years, testosterone levels remain very low (sensitive hypothalamic-pituitary axis).
   vi. At the onset of puberty there is a dramatic increase in the plasma testosterone levels. These high levels continue until age 20–25 or so, at which point they begin to slowly decline.
   vii. There is **never an abrupt decline in plasma testosterone (no male menopause).**
b. **Role in Sex Determination:** For normal male development, three mechanisms are involved:
   i. **Genetic Sex:** The sex of the child is determined by chromosome of the sperm fertilizing the ovum.
      (1) Y = Male
      (2) X = Female
   ii. **Gonadal Sex:** Do gonads become testes or ovaries?
      (1) Default system is female.
      (2) By week 6 of gestation, all embryos have an "indifferent" gonad. If the Y chromosome is present, then there is the **"SRY region", a gene** that codes for the protein **"testicular determining factor."** This causes the gonads to become testes and develop along male lines. Remember, around week 8 we see a spike and increase in androgens from the testes. **What is the problem with what we have said so far if we consider that the hypothalamus is yet to be formed?**

iii. **Phenotypic Sex:** The apparent anatomical sex.
  (1) **External genitalia:** These come from the same embryonic structures shown below:

Embryonic	Female (default)	Male
Genital tubercles	Clitoris	Glans penis
Genital fold	Labia minor	Penile shaft
Urogenital sinus	Vagina	Prostate
Genital swelling	Labia major	Scrotum

  a. The default is always female. If the embryonic structures are not exposed to DHT, they will develop female external genitalia.
  (2) **Reproductive tract:** Each embryo has immature versions of both the Wolffian system (male) and the Mullerian system (female) before differentiation.

Wolffian	Mullerian
Epididymis	Oviducts
Vas deferens	Uterus
Seminal vesicle	Upper vagina

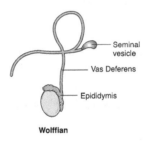

Wolffian

FIGURES 15.6

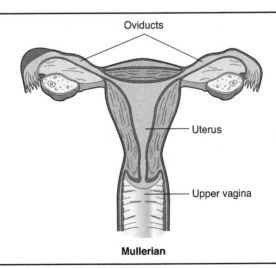

Mullerian

FIGURES 15.7

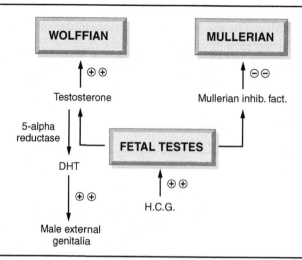

FIGURES 15.8

a. Fetal testes begin producing testosterone (due to the influence of **HCG** from the blastocyst), which stimulates the Wolffian system (causes the reproductive tract to develop along male lines).

b. The fetal testes also produce **Mullerian inhibiting factor,** which represses the Mullerian system.

c. Peripherally, testosterone is converted by 5-alpha-reductase into DHT. DHT influences the external genitalia to develop along male lines.

d. Problems: Some XY fetuses lack 5-alpha-reductase, so the gonad would develop into testes with the male reproductive tract, but the **external genitalia would be female.**

   i. **Adrenogenital syndrome:** A pathology of excess androgens. In this case, the embryo produces too many androgens. **If this were an XX embryo, what type of gonad would develop?**

     What type of reproductive tract would develop?

     What type of external genitalia would develop?

   ii. **Androgen Insensitivity Syndrome:** When androgen receptors are pathologically insensitive. This condition can vary greatly, but if in an XY embryo, it leads to no Wolffian development, no Mullerian development (still has inhibiting factor), and female external genitalia.

c. **Role in Masculinization of Brain:** Fetal estrogens cause development of the male brain.

   i. The female brain is cyclic because of the cyclic reproduction cycle. The male brain is acyclic.

   ii. The default system female (cyclic brain).

iii. The fetal testes produce androgens, of which some are converted to estrogens **(and all fetuses are exposed to maternal estrogens).**

iv. Female brains have **alpha-fetal protein**, which binds and "ties up" estrogens at the fetal brain level. Males lack this protein in significant quantities, and thus the estrogens cause the brain to develop along male lines.

d. **Role in Descent of Testes:** Testosterone plays a role in the descent of the testes.

e. **Role in Sexual Characteristics:** Testosterone remains very low until puberty. **Androgens play a role in primary and secondary sexual characteristics.**

  i. **Androgens: primary sexual characteristics**
    (1) **Enhance spermatogenesis**
    (2) **Cause growth and maturation of reproductive tract at puberty**
    (3) **Maintain reproductive tract in adulthood**
    (4) **Responsible for libido? (This is better demonstrated in nonprimates.)**

  ii. **Androgens: secondary sexual characteristics**
    (1) **Body hair: face, trunk, back, pubic regions**
    (2) **Baldness: must also have correct genotype**
    (3) **Enlarged larynx, deeper voice**
    (4) **Thickening of skin**
    (5) **Increased secretions of sebaceous glands onto skin**
    (6) **Stimulate bone growth**
    (7) **Stimulate closure of epiphyseal plates**
    (8) **Anabolic: ↑ protein synthesis**
    (9) **Increased erythropoeisis**
    (10) **Increased basal metabolic rate (BMR)**
    (11) **Increased LDLs**
    (12) **Increased synthesis of androgen-binding protein**
    (13) **Favor fat deposition (particularly in the abdomen)**

  iii. **DHT vs. Testosterone in the Fetus and Adult**

FETUS

Testosterone	DHT
Epididymis	Penis
Vas deferens	Scrotum
Seminal vesicle	Prostate
Seminiferous tubules	

ADULT

Testosterone	DHT
Penis	Scrotum
Seminal vesicle	Prostate
Sperm production	Hair (body)
Skeleton and muscle	Sebaceous glands
Libido	
Larynx	

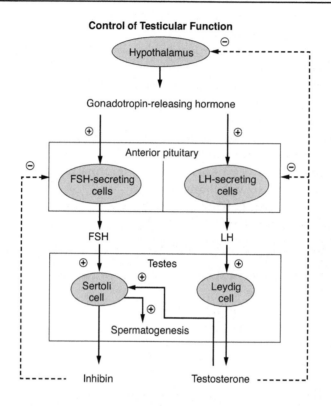

Control of Testicular Function

FIGURE 15.9

In Figure 15.9 we see a classic example of negative feedback. Notice that inhibin shuts off only at the anterior pituitary level, not at the hypothalamus.

    a.   The hypothalamus makes GnRH: This has its effect on FSH and LH (produced by the anterior pituitary).
    b.   **LH:** This goes to the Leydig cells and promotes testosterone production.
    c.   **FSH:** This causes the Sertoli cells to increase androgen binding protein and also promotes spermatogenesis.
        i.   **Spermatogenesis strongly shuts off FSH, but not LH.**
        ii.   **Inhibin** shuts off only FSH (potential birth control pill for men?).
            Could we design a birth control pill for men that shuts off GnRH? Why is this a good or poor idea?

VIII. **Age-Related Changes**
    a.   **Prepuberty to puberty:**
        i.   Testosterone levels are very low.
        ii.   There is incredible sensitivity in the hypothalamus and the pituitary.
    b.   **At the onset of puberty,** what happens in the body? We are not sure exactly what happens, but one proposed mechanism is that puberty is related to body weight and the amount of fat present in the body.

i. There is some evidence for this in that women who are too lean stop their cycles. If food is limited, then it makes no sense to make offspring.
ii. **Melatonin has also been implicated:** If you remove the pineal gland in animals, they experience an early puberty. In normal individuals, there is a decrease in melatonin secretion at the onset of puberty.
   a. Melatonin inhibits FSH and LH.
iii. **Senescence:** While males do not have the same precipitous drop as females, there is a gradual decline in testosterone. There is also a gradual decline in the **sensitivity to LH and FSH with age** (the hypothalamus and anterior pituitary are not affected, but the testes lose their sensitivity to LH and FSH).
Would a 75–year-old man have higher or lower levels of FSH and LH than you?

IX. **Xenoestrogens** (xeno = foreign): foreign estrogens
   a. There is some evidence that we are being exposed to chemicals that are estrogenic (byproduct of industrialization) or that mimic estrogens (such as petroleum, insecticides, food preservatives, and chemicals that make plastics soft). Some of the evidence (there is evidence that the exposure to estrogens is linked to abnormalities in male reproductive tract):
      i. Since 1940 there has been a **declining sperm count** (sperm per ml of ejaculate) and a **decline in the volume of ejaculate.**
      ii. Increased incidence of **testicular cancer**
      iii. Increased **reproductive tract abnormalities**
      iv. Increased **cryptorchidism** and **hypospadia** (when the penis does not form completely; urethral folds do not completely close)
      v. Increased incidence of **breast cancer** in women
      vi. This all results in **gender bending** (like hermaphroditic fish) in some species.

Female Endocrinology

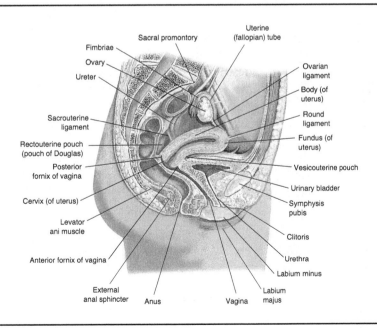

FIGURE 15.10

I. **Introduction**
   a. **Cyclic activity:** Females have periodic production of ova (eggs) and fluctuations in the sex hormone secretions.
      i. Length of cycle varies between species.
      ii. Example: **Mice have a cycle every 4–6 days. Bats cycle once a year. Humans cycle about every 28 days.**
      iii. Ovarian cycle produces cyclic changes in the lining of the uterus (endometrium). This is paramount to reproduction because it is preparing for the reception of a sperm and implantation of a zygote.
         (1) If no zygote comes together in the uterus, then the endometrium is sloughed off (only in higher primates does the cycle lead to bleeding or menstruation; it is referred to as the **menstrual cycle.** In other species it is called an **estrous cycle**).
         (2) In the estrous cycle there is no bleeding from the endometrium, but there may be **bleeding from the vagina during the time of heat (estrus).**
      iv. Fallopio was an anatomist in Italy (years ago) who was interested in when ovulation occurs in women. He sacrificed dogs to look for the presence of eggs in the oviducts.
         (1) During estrus in dogs, the vagina becomes edematous and enlarges, with a subsequent discharge (a vaginal discharge!).
         (2) Every time he found ova he found vaginal discharge. Fallopio concluded that ovulation occurs when there is vaginal discharge (true in dogs, but **in humans the discharge is sloughed-off uterus lining, not vaginal leakage**).
         (3) Fallopio's work was then used by the Roman church. He was asked for advice about avoiding pregnancies in women. As an expert, Fallopio erroneously encouraged sexual intercourse during the time that ovulation was least likely to occur—he said that the best day to avoid pregnancy was at day 14! In reality, it is just the opposite!
      v. Menarche (first menstruation): Occurs around 9–15 years of age. The cycle is active for 30–40 years (except during pregnancy), until menopause.

II. **Functional and Anatomical Overview:**
   a. **Ovaries (two of them):** They are each about the size of an unshelled almond.
      i. The outer portion of the ovary is the germinal epithelium.
      ii. The outer zone is **the cortex and contains the ova (within follicles).**
      iii. The ova in the follicles are in varying stages of development.
      iv. Inside the ovaries is the stroma.
      v. **Ovaries produce mature ova along with sex hormones.**
   b. **The Oviducts:** There is one on each side of the uterus (Figure 15.7). These have fingerlike extensions called fimbriae:
      i. The fimbriae beat back and forth and **help direct the ova into the oviduct** by creating a small current.
      ii. The oviducts extend down to the uterus and are about 4–5 inches long.
      iii. Fertilization occurs in the first third of the oviduct.

c. **The Uterus:** The uterus is **pear-shaped** and **composed of several different layers.**
   i. Its overall role is to provide an appropriate environment for the developing fetus.
   ii. It also provides a mechanical force for pushing the fetus through the cervix during parturition.
   iii. **There are three layers:**
      (1) **Epimetrium** (outside)
      (2) **Myometrium** (muscular layer)
      (3) **Endometrium** (the inner portion)
   iv. These layers are under hormonal control.
   v. The myometrium helps to **expel the fetus** during birth and is under endocrinological control (and is also **the site of menstrual cramping**).
   vi. The endometrium sloughs off during menstruation.
   vii. The cervix is usually fairly narrow. It also contains a fluid (**cervical mucous**) that is thin and watery during ovulation, yet becomes thick and viscous during pregnancy. The cervix must considerably enlarge prior to parturition.
d. **Vagina**: This is functionally a passageway for menstrual flow, childbirth, and sperm.
   i. It is a tubular, fibrous organ which is lined with **mucus secreting cells.**
   ii. There may be folds of vascularized tissue at the distal end of the vagina that partially covers the orifice. This is called **the hymen.**
      (1) The extent and durability of the hymen vary (some never have one).
      (2) The hymen may be torn due to **athletic activities, tampons, or intercourse for the first time.**
e. **Vulva:** A collective term for all the female external genitalia.
   i. **Mons pubis:** An elevation of adipose tissue covered by skin and hair that serves as a cushion for the pubic symphysis. From the mons pubis extend two longitudinal folds of skin.
   ii. **The labia: The labia majora** and the **labia minora**
   iii. **Clitoris:** A small, cylindrical mound/mass of tissue (like the penis) that engorges with blood and erects during sexual activity
   iv. **Vestibule:** A cleft space between the labia minora
      (1) Within the vestibule are the **hymen, the vaginal orifice, the external urethral orifice, and openings from several mucus-producing glands.**
      (2) The mucus-producing cells and glands provide mucus during sexual stimulation for lubrication for reception of the penis.
f. **Follicular Development:** Primordial germ cells in the embryo first produce oogonia. These undergo mitosis until about week 20–24 of fetal development.
   i. At week 20–24 of fetal development in the female, there are about 7 million oogonia. Until about 6 months after parturition, these oogonia undergo prophase of the first meiotic division (and become **primary oocytes**).
   ii. After the first prophase of the first meiotic division, **the process is arrested** and you have a primary oocyte. Some of these oocytes will remain suspended for many years (until menopause).
   iii. The next step is the conversion of the naked oocyte into a **primordial follicle** (by laying down of a single layer of granulosa cells around the primary oocyte).
   iv. During this time, a significant amount of attrition takes place. At about 6 months of age, there are only about **2 million primordial follicles.**
   v. The 2 million primordial follicles have one of two fates:
      (1) To **become a mature follicle and ovulate**
      (2) To **become atretic** (never develop)

    vi. Before puberty, many primordial follicles begin to develop and may reach the primary follicle state, but from there they all will become atretic. **Only about 400,000 ova are still viable (out of the 7 million) by puberty.**

    vii. From puberty to menopause, various follicles begin to develop. These follicles will become mature and ovulate only if the hormonal environment is ideal.

    viii. Out of the 400,000, only **about 400 will ovulate**. The rest will become atretic.

    ix. At the time of birth, females have all the ova that they are ever going to have.

        (1) Some of the ova are in meiotic arrest during birth and may not ovulate for 40 years.

III. **The Ovarian Cycle:** The ovarian cycle is everything that happens in the 28 days. There is a follicular phase (days 1–14) and a luteal phase (days 15–28).

  a. **Follicular Phase**

    i. The primordial follicle (oocyte in the middle, with a single layer of granulosa cells around it) grows and the granulosa cells proliferate (more than one layer). The **zona pellucida (layer between the ova and the granulosa cells)** develops now. This is a hormonally independent action.

    ii. **From the primary follicle to the secondary follicle:** Under the influence of **FSH**, we get additional proliferation of granulosa cells and a new, outer layer develops—the **thecal layer.**

    iii. The thecal layer and the granulosa layer **begin to produce estrogens**.

        (1) It is the secondary follicle that begins to show a cavity (the **antrum**). The antrum is filled with **estrogens, electrolytes, and plasminogen.**

    iv. **From the secondary follicle to the tertiary follicle:** There is one dominant tertiary follicle that develops over the others. The other follicles then begin to shrink back and become atretic.

        (1) The one tertiary follicle continues its secretion of estrogen.

        (2) The thecal cells are developed into the **theca interna** and **theca externa**.

        (3) The **antrum gets larger**.

    v. Finally, we get a **mature follicle** (Graafian follicle) with the granulosa cells, the theca interna, and the theca externa. This follicle is now destined to ovulate.

    vi. At the time of ovulation, there is a surge of LH. The LH can bind receptors on the granulosa cell layer. This LH surge signals the ova to complete meiotic division and causes the release of plasminogen activating factor (which makes plasmin).

        (1) Plasmin causes the follicle to burst and the ova to burst out into the abdominal cavity, where the fimbriae help direct it into the oviduct.

    vii. All of the above happens in the follicular phase (between days 1 and 14 of the cycle). Once the follicle ruptures, this signals the end of the follicular phase and the beginning of the luteal phase.

  b. **Luteal Phase:** After ovulation, the ruptured follicle undergoes biological changes and is transformed into the corpus luteum. **The corpus luteum produces estrogen and progesterone for several days (10 or so) beyond ovulation, unless pregnancy occurs (in which case it is maintained).**

    i. If there is no pregnancy, the corpus luteum becomes less active and degenerates. It finally degenerates enough so that it no longer produces estrogen or progesterone and becomes the **corpus albicans.**

    ii. The granulosa and thecal cells form the **corpus luteum**, which then hypertrophies and finally degenerates into the **corpus albicans.**

IV. **The Endometrial Cycle:** Composed of three phases
  a. **The menstrual phase:** The sloughing off of the endometrium (Days 1–5)
  b. **The proliferative phase (or the estrogen demand phase)** (Days 6–14): This is primarily influenced by the developing follicle. The endometrium is getting thicker due to estrogens.
  c. **The secretory phase (or the progesterone demand phase)** (Days 15–28): The endometrium is lush and edematous and ready for implantation.

V. **Cervical Mucus Cycle:** There is mucus within the cervical neck that goes through a cycle which is dependent upon hormones.
  a. **Estrogen phase:** The mucus is thin, watery, and alkaline.
    i. This is advantageous for the sperm to travel to the oviducts to fertilize the ovum.
    ii. This is estrogen dependent.
    iii. This occurs during the middle of the cycle.
  b. **Luteal phase:** There is a thick, viscous, acidic mucus.
    i. This is progesterone dependent.

VI. **Breast cycle**
  a. **Follicular phase—estrogens promote:**
    i. Deposition of adipose tissue
    ii. Accumulation of fluid in the breasts
    iii. Increased size of nipple
    iv. Increased pigmentation of nipple
  b. **Progesterone phase:** Progesterone sustains the above effects until menstruation occurs. After menstruation, the tissues return to their baseline values.
  c. **In order for lactation to occur, pregnancy hormones are needed!**

VII. Function of Estrogens: Estradiol 17-β is the most significant in the nonpregnant female.
  a. **Primary Sexual Characteristics:** Estrogen
    i. Causes the growth and maintenance of the female reproductive tract
    ii. Increases vaginal thickness in the epithelium so the vagina is not damaged during sex
    iii. Causes endometrial proliferation
    iv. Increases myometrium excitability
    v. Increases the motility of the oviducts
    vi. Causes the deposition of fat in the breasts
    vii. Stimulates the ductile system development of the breasts
    viii. Causes the thinning of the cervical mucus
    ix. Causes the formation of LH receptors in the granulosa cells
  b. **Secondary Sexual Characteristics:** Estrogen
    i. **Affects the skeletal system**
      (1) Influences the shape of the pelvis and favors bone deposition
      (2) When women go through menopause (little or no estrogen), the bones become thin and weak and osteoporosis may set in.
      (3) Also favors the closure of the epiphyseal plates, more so than androgens do. This is one reason why males are taller than females.
    ii. **Influences Muscle:** Estrogen favors protein deposition (but not as much as androgens do).
    iii. Increases the **fat deposition** beneath the skin at the thighs and hips
    iv. Causes the **softening of the skin** and increased **vascularization of the skin**
    v. Causes an **increase in HDLs** and a **decrease in LDLs**

      vi. Causes **increased production of clotting factors**
          (1) This was a big issue with the first birth control pills, which caused thromboemboli (because the pills contained a lot of estrogen).
     vii. There is **not much effect on hair deposition,** which is mostly due to androgens from the adrenal gland.
    viii. Libido?

VIII. **Functions of Progesterone**—Progesterone has several effects on the:
    a. **Uterus:** Progressively increases the secretory capacity of the endometrium and causes the myometrium to be less active
    b. **Breasts:** Responsible for the development of the lobules and the alveoli
    c. **Cervix:** Causes a thickening of the cervical mucus
    d. **Electrolytes:** Progressively causes a decrease in Na^+ retention
    e. Causes an increase in **body temperature**

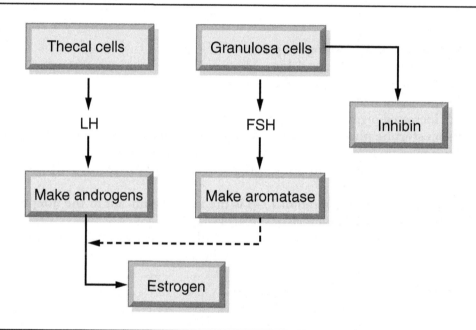

FIGURE 15.11

The Female Reproductive Tract

IX. **Regulation of Female Cycle**
    a. The early stages (primordial to the primary follicle) are not dependent on hormones.
      i. For proliferation of the granulosa cells and the thecal cells, there must be an **increase in FSH**. This will then cause the follicle to develop into a secondary follicle (with a well-differentiated thecal and granulosa cell layer).
     ii. Estrogen begins its steady climb.
    iii. Must have thecal cells and granulosa cells cooperating to make estrogen

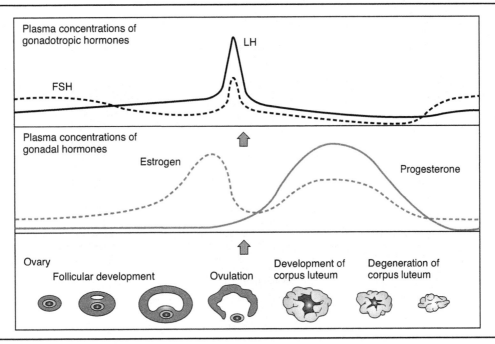

FIGURE 15.12

iv. As the estrogen concentration increases, there is an effect on the endometrium (proliferation of the endometrium). **As estrogen levels increase, we also see an effect on FSH; this negative feedback on FSH serves to then decrease FSH levels.**

v. **Estrogen is feeding back to the anterior pituitary to inhibit FSH.**

vi. Granulosa cells also produce **inhibin, which directly inhibits FSH production.**

vii. During this time, **estrogen levels reach some appropriate level with a little bit of progesterone, a little bit of inhibin, and perhaps less activin (let's call this "cocktail #1").**
   (1) With cocktail #1, the hypothalamic anterior pituitary flips over to a **positive feedback mode.**
   (2) Now estrogen causes more production of FSH and LH and you see a **spike of LH and FSH.**
   (3) The spike of LH is essential for ovulation to occur (it causes the activation of plasmin via plasminogen activating factor, which causes the rupture of the follicle on day 14 and subsequent ovulation).
   (4) The corpus luteum is then formed from the thecal and granulosa cells and it then begins to produce estrogen (and lots of progesterone).

viii. We can see an initial **drop in estrogen due to the reconfiguring of the follicle.**

ix. **"Cocktail #2"** (high estrogen, low inhibin, and very high progesterone levels) then causes a flip back into the **negative feedback mode.**

x. FSH and LH levels begin to go down because of the large amounts of progesterone and estrogen, which inhibit LH and FSH.
   (1) Eventually, the corpus luteum degenerates (so estrogen and progesterone levels plummet) into the **corpus albicans.**
   (2) The endometrium is sloughed off and FSH and LH gradually go up again as a new cycle begins.

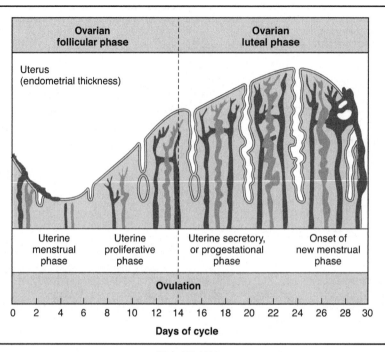

Ovarian follicular phase | **Ovarian luteal phase**

Uterus (endometrial thickness)

| Uterine menstrual phase | Uterine proliferative phase | Uterine secretory, or progestational phase | Onset of new menstrual phase |

Ovulation

Days of cycle
0 2 4 6 8 10 12 14 16 18 20 22 24 26 28 30

FIGURE 15.13

X. Birth Control
a. Inhibits Ovulation
    i. **Birth control pill:** Contains a synthetic estrogen-like substance and synthetic progesterone (the liver breaks down real estrogen and progesterone quickly).
        (1) Take for 3 weeks and then take a placebo for 1 week.
        (2) Increase estrogen and progesterone levels
        (3) GnRH levels decrease (and stay low).
        (4) FSH and LH stay low so that normal follicular development does not occur.
        (5) There is still a development of the endometrium, but when the hormones are stopped, it sloughs off.
        (6) The hormones keep the cervical plug viscous so that sperm can not get through.
        (7) Patient has to be vigilant for the birth control to work, so compliance can be a problem.
        (8) Example: (NORPLANT)—synthetic progesterone (implanted under the skin).
b. Barrier
    i. **Condom:** Latex sheath that provides a barrier between the sperm and the ova. Generally, this is not as effective as the pill but is still fairly effective.
        (1) There are varying efficacies: proportion of pregnancies per year per 100 normally sexually active women is 5%–10%. User efficacy is a concern.
    ii. **Diaphragm:** Device inserted into the vagina, near the cervix.
    iii. **Cervical caps, vaginal pouches, contraceptive sponges.**
    iv. **Minipill:** A progesterone-only preparation that thickens the cervical mucus to prevent sperm penetration.
c. Chemicals: Products (that contain spermicide) such as foams, creams, and jellies that are put in the vagina. Usually these are used in combination with barrier devices.
    i. **Nonoxynol-9:** The active ingredient in most spermicides.
        (1) Efficacies vary but are better if a barrier is used with it.

d. **Sterilization**
   i. **Vasectomy:** Cut and tie the vas deferens surgically. This creates a sperm-free ejaculate.
   ii. **Tubuligation:** Tie off the oviducts surgically so that the ova cannot meet the sperm in the oviducts.
e. **Prevent Implantation**
   i. **IUD: Intra-Uterine Device**
      (1) Copper is inserted into the uterus by a physician while the cervix is dilated.
      (2) It disturbs the environment of the uterus.
      (3) There are some problems with pregnancies and bleeding.
      (4) IUDs raise moral issues because they prevent implantation, not fertilization.
      (5) **(MIRENA)**—an IUD that also delivers a synthetic progesterone so as to also prevent fertilization by causing a viscous cervical mucus. This helps overcome possible moral objections to original IUDs.
   ii. **Morning-After Pill:** A high estrogen dose to increase the motility of the oviducts and the uterus.
      (1) They hasten the transport of the zygote through the reproductive tract (no implantation occurs).
   iii. **RU 486 (mifepristone):** A progesterone antagonist that increases the motility of the uterus and displaces the blastocyst before implantation (day 8).
   iv. **Prostaglandins:** Increase motility; prevent implantation
f. **Avoid Fertile Period**
   i. **Rhythm Method:** Failure rate of 20%–30% (pregnancies per 100 women).
      (1) Because fertilization takes place when the sperm and the egg meet, avoid sex during ovulation (day 14).
      (2) May not be very successful because ovulation time varies.

XI. **Pharmacology**
   a. **Estradiol (ESTRACE)**
      i. Prescribed predominantly for post-menopausal women to prevent the symptoms of menopause
      ii. A therapy for osteoporosis
      iii. Not naturally occurring estrogens (these are synthetic)
   b. **Conjugated estrogens (PREMARIN)**
      i. Naturally occurring estrogen from pregnant mares
   c. **Progestin (PROVERA)**
      i. Synthetic progesterone
      ii. Given for abnormal bleeding during cycles
   d. **Combination of estrogen/progesterone (PREMPRO)**
      i. For menopausal symptoms, estrogen alone is sometimes not sufficient
      ii. May also be used to treat prostate cancer in males
      iii. **Estrogen taken alone increases chance of endometrial carcinoma**
      iv. Endometrial carcinoma chances decrease if progesterone is added.

XII. Implantation

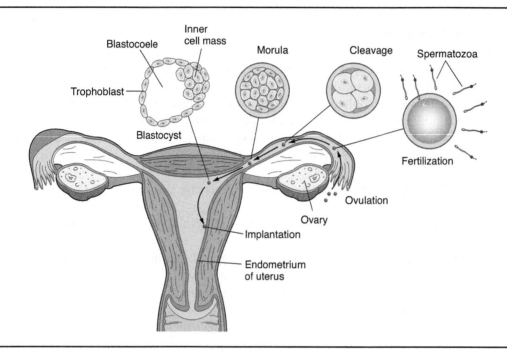

Blastocoele

Inner cell mass

Morula

Cleavage

Spermatozoa

Trophoblast

Blastocyst

Fertilization

Ovulation

Ovary

Implantation

Endometrium of uterus

FIGURE 15.14

a. **Hormones**
   i. After intercourse the sperm reaches the ends of the oviducts in about 5 minutes (too fast to just be swimming!).
   ii. Proposed reason for getting there so quickly:
      (1) Prostaglandins in the semen increase the motility of the myometrium and force the sperm through.
      (2) Oxytocin increases motility of the myometrium and uterus.
b. **Sperm and Egg Coming Together**
   i. Sperm are viable in the reproductive tract for 2–3 days but are maximally viable for only about 24 hours.
   ii. 2–6 million sperm are deposited in the vagina, but only about 200 reach the vicinity of the ovum.
   iii. Many sperm are needed to prepare the ovum.
   iv. Only 1 sperm fertilizes the ovum. This usually occurs at the first third of the oviduct.
   v. Fimbriae are active to direct the ova into the oviduct.
   vi. The ova are viable for 12–24 hours but are maximally viable for only about 8–12 hours.
   vii. 2–3 days are required for the zygote to go down the rest of the oviduct and reach the uterus. Another 3–4 days are needed for implantation.
   viii. Implantation occurs in the properly prepared endometrium.
   ix. Basically, the cells of the blastocyst (trophoblastic cells) eat a hole in the endometrium and then attach to it.

XIII. Hormones of Pregnancy

a. **Human Chorionic Gonadotropin (HCG)**

    i. **Days of Cycle:** 1–5 is menstruation, 5–14 is proliferation, and 14 is ovulation.

    ii. It takes 3 days to get out of the oviduct (so about day 17) and then another 3 to 4 days to implant.

    iii. At day 22, the corpus luteum decreases to the corpus albicans. The endometrium is sloughed off.

    iv. Something has to interrupt the cycle for pregnancy to occur.

    v. Initially, the trophoblastic cells of the blastocyst produce HCG.

    vi. **HCG functions**

        (1) In the mother: HCG is very similar to LH (in structure and biochemistry). The corpus luteum gets bigger and continues producing estrogen and progesterone through the first trimester of pregnancy (until the placenta can take over) to keep the endometrium maintained. **HCG maintains the corpus luteum.**

            a. The HCG hormone is detected in pregnancy test kits. Only pregnant females produce HCG. Early on, it may not be dictated in the urine tests, but the blood tests will detect HCG within one week. Male urine may test positive for HCG if testicular cancer is present.

        (2) In male fetus: Causes the embryonic testes to produce testosterone. Adult testes produce testosterone because of LH (an 8–week-old fetus does not have an anterior pituitary, so HCG must stimulate the secretion of testosterone from the testes).

b. **Estrogens:** The estrogens of pregnancy

    i. During the 1st trimester the estrogen is produced predominantly in the corpus luteum

        (1) During the 2nd and 3rd trimesters: estrogen produced in the placenta.

        (2) Estrogen is not produced in the placenta in the 1st trimester because the placenta is underdeveloped and not capable of producing estrogen all the way from cholesterol.

    ii. **The 2nd to 3rd trimester: The fetal placental unit**

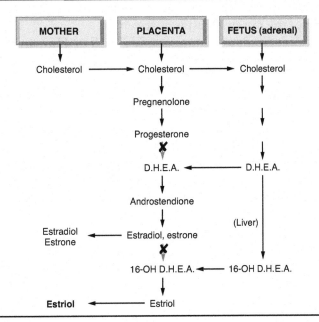

FIGURE 15.15

(1) Fetus provides the raw material for producing estrogen. This cannot happen in the 1st trimester because fetal adrenal glands are not sufficiently developed.

(2) Estriol is the most common estrogen in pregnant females.

(3) Follow Figure 15.15, and **notice where Xs are; these denote transitions that cannot be made, and thus the fetal adrenal glands must provide substrate.**

   iii. **There has to be cooperation between the placenta and the fetus.**

(1) Even with just the placenta, estrogen could not be produced because there are no fetal adrenal glands present.

(2) During pregnancy, the estrogen most responsible for the estrogen effects is estriol.

   iv. **Special functions of estriol**

(1) Enlargement of the external genitalia and uterus

(2) Enlargement of the breasts

(3) Specific for the growth of the ductile system

(4) Relaxation of the pubic ligaments

Is it a coincidence that many miscarriages happen around weeks 10, 11, or 12, the same time that there is a switch from "corpus luteum–produced estrogen" to "placental estrogen"?

  c. **Placental Progesterone**

   i. 1st trimester: From corpus luteum

(1) 2nd and 3rd trimester: From placenta

   ii. **Functions:**

(1) The decidual cells (special cells of the uterus) are induced to produce nutrients for the free floating zygote.

(2) Progesterone is a substrate for the fetus to produce cortisol and aldosterone.

(3) Decreases myometrial activity

(4) Inhibits oxytocin sensitivity

(5) Decreases prostaglandin activity (all work to maintain pregnancy)

(6) Causes the development of the alveolar pouches in breast

(7) Causes the thickening of the cervical plug

(8) Stimulates the maternal respiratory center

  d. **Human Chorionic Somatomammotropin (HCS or Human Placental Lactogen**): Produced by the placenta. HCS levels are used to assess the growth and health of the fetus. **Functions:**

   i. Prepares the breasts for lactation and stimulates the mammary glands

   ii. Increases fetal somatomedins (IGFs)

(1) In the fetus, there is high hGH, but hGH has no effect on fetus growth. Anencephalic (no brain) neonates are normal size.

   iii. **Decreases the insulin sensitivity in the mother** (this benefits the fetus). This may cause pregnancy-induced diabetes. HCS is working against insulin.

  e. **Relaxin:** Produced by the corpus luteum and then by the decidual cells of the endometrium. **Functions:**

   i. Causes the **relaxation of the pelvic floor**

   ii. Causes the **softening of the cervix**

   iii. Reduces the **myometrial contractions (prepares the pelvis and cervix for parturition)**

XIV. Parturition: The length of gestation varies considerably. In humans it is 38 weeks (elephants: 20 months; dogs/cats: 9 weeks). No one knows for sure what happens. In sheep, the mechanism is well known and the fetus signals the onset of parturition. In humans, it is more complex (hormonal and mechanical factors).

  a. **Hormonal Factors:** Progesterone and estrogen ratios
    i. **Progesterone** *decreases* uterine motility, and **estrogen** *increases* uterine motility.
      (1) In many species you see a change in these concentrations.
      (2) At the end of pregnancy, there is an increase in estrogen and a decrease in progesterone so that there is an increase in motility of the uterus.
      (3) In humans there is no change in estrogen and progesterone concentrations in the plasma.
      (4) Locally, the production of estrogen and progesterone is by the membranes of pregnancy (amnion and chorion). Perhaps these levels change? These may never be reflected in plasma levels.
    ii. **Oxytocin:** Released in response to the stretching of the cervix
      (1) Oxytocin hastens parturition in a uterus that is properly prepared.
      (2) Oxytocin levels do not rise until after the onset of parturition.
      (3) Changing the estrogen and progesterone levels can change oxytocin receptor sensitivity (estrogen *increases* receptor sensitivity and progesterone *decreases* receptor sensitivity).
      (4) Recent evidence: **Uterus also synthesizes oxytocin, and this is not reflected in the plasma concentration.**
      (5) The levels in the plasma do not go up until after the onset of parturition. (**PITOCIN**) is a synthetic oxytocin used to induce labor in the properly prepared uterus.
    iii. **Prostaglandins**
      (1) Increase uterine motility
      (2) The exact role in the initiation of parturition is unknown.
      (3) What do prostaglandins *not* do?
    iv. **Fetal Endocrine Signal (perhaps it is signal from fetus, *not* maternal)**
      (1) In sheep: Fetal pituitary releases ACTH, which causes the production of **cortisol from the fetal adrenal glands**, which increases estrogen synthesis and prostaglandins in the mother.
      (2) Cortisol does other things in the fetus: Assists in lung maturation, closes ductus arteriosis (shunt between the pulmonary artery and the aorta), and assists in the development of intestinal transport mechanisms.
      (3) If you inject ACTH into sheep, they prematurely parturate.
      (4) In humans, it is more complicated: ACTH and glucocorticoids do not induce parturition.
      (5) In humans, **anencephalic fetuses have longer gestation (as do those with adrenal hypoplasia)**, so there seems to be some effect.
  b. **Mechanical Factors**
    i. As the fetus develops, it crowds the uterus and causes it to stretch. Toward the end of the pregnancy the baby "drops" (is engaged with its head into the cervix).
      (1) The head turns downward and stretches the cervix, which causes the myometrium to contract and pushes the fetus more downward.
      (2) When the myometrium contracts again it pushes the fetus more downward and so on…
      (3) Women may experience "Braxton Hicks contractions" during this time (not the real thing).
      (4) Stretching the cervix also induces oxytocin release.
      (5) You can induce labor manually (breaking membranes, etc.).

XV. **Lactation:** Mammary glands are present in both sexes.

Mammary Gland Anatomy

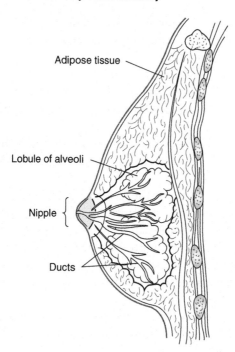

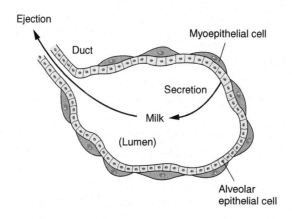

FIGURE 15.16

a. **Gynecomastia:** Males develop breasts if estrogens get expressed.
b. Breasts develop in females because of estrogen and progesterone.
   i. Each gland has around 20 **lobules**. Each of these radiate via the ductile tissue to the nipples.
   ii. Each lobule has a number of **alveoli** (functional units of the breast) that produce milk.
      (1) Alveolus: Glandlike structure that produces milk
      (2) In nonpregnant females, the milk production apparatus is rudimentary and there is very little ductile development.
      (3) Typically an alveolus has a secretory epithelium layer and a myoepithelial layer (see above figure).

c. **Hormones**
   i. **Estrogen responsible for:**
      (1) Growth of the breast
      (2) Deposition of fat in breast
      (3) Development of the ductile system
   ii. **Progesterone**
      (1) Causes cells to become secretory in nature in pregnant women.
      (2) Increases the development of the alveoli.
   iii. **Prolactin**
      (1) Essential for the secretion of milk from the epithelial cells into the lumen (progesterone and estrogen inhibit).
   iv. **HCS (HPL):** Similar to prolactin but not as effective. This increases the density of prolactin receptors.
      (1) By the end of pregnancy the breast is ready for milk production, but there is no secretion of milk into the alveoli without prolactin (needed to get secretion).
      (2) Estrogen and progesterone inhibit prolactin (the source of estrogen and progesterone is the placenta).
      (3) At parturition, the placenta is expelled from the uterus. Now, prolactin can be secreted and the milk can be released into the alveoli. The milk still has to be let down to the level of the nipple, though.
   v. **Oxytocin** is released from the posterior pituitary in response to sucking by the infant. Functions:
      (1) Tells myoepithelial cells to contract, and milk is let down into the ductile system and to the nipple, where it is available to the infant.
         a. **Colostrum:** This is what we call the first milk from the nipple that is lower in fat and lactose than true milk. It has an increased level of proteins (immunoglobulins). Several hours before milk is available: If mothers are going to breast feed, nurses usually bring the baby in soon after parturition to nurse even though nothing much is happening. There is not much milk in the alveoli, but the secretion of oxytocin reduces the amount of bleeding in the uterus (the suckling baby causes the mother to release oxytocin). This ultimately causes the myometrium to contract down and thus starts restoring the size of the uterus.

XVI. **Pathologies**
   a. **AMI (Athletic Menstrual Irregularity):** A number of female athletes have irregular menstrual periods (stopped or infrequent cycles).
   Lack of menstruation = amenorrhea
   Infrequent menstruation = oligomenorrhea
   No ovulation, but still have cycle = inovulatory
   Painful menstruation = dysmenorrhea
   i. 5%–50% of women report AMI
   ii. Study: 28 women (college age), athletic but not in training (with normal cycles), were exposed in the first 4 weeks to running (4 miles a day). For the next 5–8 weeks they ran 10 miles a day. At the end of the study, only 4 had normal menstrual cycles. All of the women returned to their normal menstrual cycles after 6 months.
   iii. There is decreased progesterone in the luteal phase and decreased estrogen in the follicular phase.
   iv. AMI may have something to do with % body fat.
   v. **Women functionally become menopausal** and have the risks of osteoporosis.

b. **Endometriosis**
   i. Functional endometrial tissue where it does not belong (ovaries, outside the uterus, uterine ligaments, vagina, intestines; and in bizarre places like the nostril, lungs, or limbs).
   ii. Etiology is unknown.
   iii. **Theories**
      (1) **Retrograde Menstrual Theory:** During menstruation some flow finds its way back to the oviducts to begin functioning. But how does this explain endometrial tissue locations such as nostrils?
      (2) **Embryogenic Implantation Theory:** Undeveloped endometrial tissue gets misplaced during embryological development and later becomes active.
      (3) **Endometrial Metastasis:** Tissue gets in the lymphatic or circulatory system and is transported to cells elsewhere. This is much like cancer cell metastasis.
   iv. Patients typically undergo menstruation and the sloughing off cycle, but it causes severe pain, scarring, adhesions, and sometimes even sterility (blocked oviducts, so there is no fertilization).
   v. **Demographic data:** Incidence is higher in higher socioeconomic levels, women between ages 20 and 40, women in developed countries, and goal-oriented women (overachievers or type A personalities).
   vi. **Treatment:** Treatment with contraceptives (birth control) usually relieves the symptoms. Also:
      (1) **Drug (DANAZOL):** Synthesized androgen that suppresses GnRH, FSH, and LH so that a normal cycle does not ensue. The side effects are menopause and mascularization (hirsutism = facial hair).
      (2) **Surgery:** May laparoscopically cut tissue away or perform surgery such as hysterectomy or ovariectomy
c. **Ectopic pregnancy:** Fertilized ovum implants somewhere else other than the uterus (in the oviducts is most common). Such pregnancies are destined for failure because the fetus outgrows the space.
d. **Dysmenorrhea:** Painful menstruation
   i. Prostaglandins have been implicated.
   ii. Treatment is to use a prostaglandin inhibitor (taken before menstruation occurs).
      (1) **Example: (MOTRIN), (NAPROSYN), (IBUPROFEN)**
e. **PMS: A distinct clinical phenomenon**
   i. Physiological and psychological symptoms
   ii. Occurs during the premenstrual period (10 days or less)
   iii. Etiology is unknown
   iv. More common in females in their 30s than in teens and 20s
   v. Bloating, abdominal pain, swelling breasts, depression, headaches
   vi. Treatment multifactorial: dietary changes, increased exercise
   vii. 30%–40% of women report some symptoms
f. **Osteoporosis**
   i. **Pathology:** Typically present in postmenopausal women. Because estrogen promotes bone density, **lack of estrogen can lead to this condition, characterized by porous and brittle bones.**
   ii. **Pharmacology**
      (1) **(EVISTA):** Stimulates estrogen receptor on bone. This has an estrogen-like effect on bones.
      (2) **(FOSAMAX):** Inhibits bone resorption (osteoclastic activity)

(3) (MIACALCIN): This drug is salmon calcitonin. In nonphysiological doses, this works to decrease Ca^{++} resorption from bone. Usually this is administered as a nasal spray.

SUGGESTED READINGS

1. Guyton, A.C., and Hall, J.E. 2011. Chapters 80–82 in *Textbook of Medical Physiology*, 12th ed. Baltimore: Saunders.
2. Koeppen, B.M. and Stanton, B.A. 2008. Chapter 43 in *Berne & Levy Physiology*, 6th ed. St. Louis: Mosby.

CHAPTER 15 QUESTIONS

1. _____ A component of the vulva
   A. Clitoris
   B. Vagina
   C. Mons pubis
   D. All of these
   E. Two of these

2. _____ Likely to cause a decrease in androgen production
   A. Cryptorchidism
   B. Inguinal hernia
   C. Vericocele
   D. None of these
   E. Two of these

3. _____ The genital folds
   A. Are found in both sexes
   B. Are inhibited by Mullerian inhibiting factor
   C. Differentiate into the shaft of the penis in response to testosterone
   D. All of these
   E. Two of these

4. _____ Sertoli cells
   A. Produce androgens
   B. Produce inhibin
   C. Form a blood testis barrier
   D. All of these
   E. Two of these

5. _____ Would likely be found in a 6-year-old girl's ovaries
   A. Corpora albicans
   B. Primordial follicles
   C. Estrogen
   D. None of these
   E. Two of these

6. _____ Produced by the ovary
   A. Progesterone
   B. Plasminogen
   C. Inhibin
   D. All of these
   E. Two of these

7. _____ Progesterone
   A. Is produced during the follicular phase of the ovarian cycle
   B. Is a steroid
   C. Decreases the activity of the myometrium
   D. All of these
   E. Two of these

8. _____ Causes the fetal testes to secrete androgens
   A. FSH
   B. LH
   C. Testosterone
   D. HPL
   E. None of these

9. _____ Oxytocin
   A. Is secreted into the primary plexus
   B. Causes the secretion of milk
   C. Increases the contractions of the endometrium
   D. None of these
   E. Two of these

For questions 10–15, choose one of the following:
   A. Increases or is greater than
   B. Decreases or is less than
   C. Has no effect or is equal to

10. _____ The effect of estrogens on the thickness of the cervical mucus

11. _____ Plasma androgen levels in a 4-month-old fetus, as compared to a 4-year-old child

12. _____ The effect of androgens on LDLs

13. _____ The effect of estrogens on the formation of the clotting factors

14. _____ The number of corpora lutea, as compared to the number of atretic follicles

CHAPTER 16

Endocrinology IV: The Pancreas

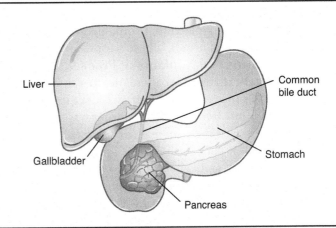

Figure 16.1

I. **Exocrine Portion of the Pancreas:** Has digestive functions
   a. **Acinar cells produce pancreatic juices and bicarbonate.**
   b. **Primary and Secondary Pancreatic Ducts**
      i. Pancreatic enzymes and HCO_3^- are secreted by the primary and secondary pancreatic ducts into the duodenum of the SI (all people have primary duct—some people have secondary duct).
   c. **Hormonal Regulation:** The secretions are governed by neuronal and hormonal mechanisms. There are two hormones that do this.
      i. **Secretin:** The presence of acid chyme in the SI causes the intestinal mucosal cells to release secretin into the bloodstream.
         (1) Secretin circulates back to the pancreas and causes a bicarbonate-rich secretion to be released.
      ii. **Cholecystokinin:** Released due to the presence of amino acids and small peptides in the SI, but particularly in response to partially digested fats.

(1) The intestinal epithelial cells secrete this into the blood. It then travels back to the pancreas and causes the secretion of an enzyme-rich substance.

II. Endocrine Portion of the Pancreas: Has metabolic functions
a. Introduction
  i. Cell Types: The Islets of Langerhans—There are many islets within the pancreas. There are several different cell types.
  - **(1) Alpha cells (α):** (20%–25%) produce glucagon
  - **(2) Beta cells (β):** (60%–70%) produce insulin
  - **(3) Delta cells (δ):** (10%) produce somatostatin (GHIH) and a small amount of gastrin (this is not the primary site for gastrin production)
  - **(4) F-cells:** (very few) produce pancreatic polypeptide

  ii. Embryology: Islet cells develop before the acinar cells.
  - **(1)** Acinar cells are involved in digestion; the fetus does not need a digestive system until parturition.

  iii. History of Insulin
  - **(1) Aristotle:** Recognized that some people urinated quite a bit (and that flies and ants were attracted to this urine). These people usually died due to diabetes mellitus.
  - **(2) Von Mering and Minkowski:** Produced diabetic dogs by taking their pancreas out.
  - **(3) Schultze:** Ligated the pancreatic duct in dogs, and they did not exhibit any diabetic symptoms (because the pancreatic duct is exocrine, not endocrine).
  - **(4) Banting and Best:** (1920s) Banting was a professor and Best was his graduate student. They removed the pancreas (or other organ) from dogs and would "grind" the organ up and put it back in the dogs. They would observe dogs after removal to see if they were affected and, when they put the organ back, if the dogs were relieved. When they removed the thyroid gland, ground it up, and then put it back in, the symptoms were relieved. This did not work with the pancreas. **Best tried an experiment on a fetal pig:** He removed the pig pancreas and then ground it up and put it in the dog. Subsequently, the dog's symptoms were relieved (but he had to keep injecting with pig pancreas for symptom relief to continue). Acinar cells were not yet present in the fetal pig (they produce digestive enzymes). Thus, the dog was receiving only the endocrine portion. In the dogs, the exocrine portion of the ground-up pancreas was digesting the insulin before it could make a difference. **A Nobel Prize was awarded to Banting and McLeod, not Best (although Banting split his prize money with Best).**

b. Insulin: A protein
  i. Synthesis: In the β cells of Islets
  - **(1)** Synthesis begins in the beta cells at ribosomes. What is synthesized is actually proinsulin, which is taken into the Golgi apparatus and then converted into insulin (the real thing).
  - **(2) Remember: Insulin secretion is stimulated under circumstances of fuel excess, and inhibited under circumstances of fuel deficiency.**

  ii. Secretion: Factors influencing insulin secretion
  - **(1) Elevated plasma glucose:** This is the most profound stimulus. Insulin is released rapidly, within a few minutes, from stored insulin. There is a decline and then a steady climb because new insulin is being synthesized. **This causes a biphasic response by the pancreas.**

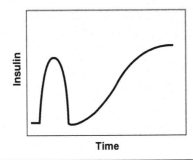

FIGURE 16.2

(2) **Elevated plasma amino acids:** Increased plasma amino acids are potent stimulators of insulin secretion.

(3) **Elevated Fatty Acids:** Stimulators of insulin secretion

(4) **G. I. Hormones:** Glucagon-like peptide 1 (GLP-1) and Gastric Inhibiting Peptide are both typically secreted in response to food in the gut. **Both cause an increase in insulin secretion and a decrease in GI motility.**

(5) **Neural Regulation by the ANS:** Parasympathetic control *increases* insulin secretion, while sympathetic control *decreases* insulin secretion.

(6) **Somatostatin (GHIH):** Also produced by the GI system, this inhibits insulin secretion.

iii. **Actions**

(1) **Carbohydrate (CHO) metabolism**

   a. Increases glucose uptake into cells (by GluT4 translocation from the inside of the cell to the cell membrane)

   b. Manufactures additional transport molecules

   c. Stimulates the activity of glucokinase (hexokinase IV). Thus, in the liver, we are able to create a large gradient by activating this and phosphorylating glucose as soon as it enters the cell. This large gradient is needed for diffusion into the cell.

   d. Stimulates glycogenesis in liver and skeletal muscle

   e. Inhibits glycogenolysis

   f. Inhibits gluconeogenesis

     (With exercise/contractions, GluT-mediated transport into cells is independent of insulin.)

(2) **Fat metabolism: Hormone-sensitive lipase breaks fats down inside of cells (to then export them), while lipoprotein lipase breaks down fats outside of cells (usually to import them).**

   a. Increased production and storage of fat

   b. Increased transport of glucose into adipocytes

   c. Converts glucose derivatives into FAs

   d. Increases transport of FAs into adipocytes

   e. Inhibits hormone-sensitive lipase (hydrolyzes fats)

   f. Inhibits β-oxidation

(3) **Protein metabolism: Acts much like hGH in terms of proteins. Unlike hGH, it decreases plasma glucose.**

   a. Enhances ribosomal protein synthesis

   b. Enhances active transport of amino acids into cell

   c. Decreases protein catabolism

(4) Other actions
    a. Increases the movement of K^+ from the extracellular fluid to the intracellular fluid
    b. Decreases the activity of neuropeptide y (produced in the brain, this peptide stimulates hunger)

c. Glucagon: Produced by the alpha cells (and the glucagon-like peptides 1 and 2 are produced by the intestinal mucosa). An antagonist to insulin. The major site of action is the liver (predominant effect on the liver).
    i. Actions:
      (1) On Carbohydrates
        a. Increases glycogenolysis
        b. Increases gluconeogenesis
      (2) On Fats
        a. Increases lipolysis
        b. Increases β-oxidation
      (3) On Proteins
        a. Increases the uptake of gluconeogenic precursors in the liver
    ii. Regulation of glucagon
      (1) Glucose: Decreased plasma glucose concentration stimulates secretion
      (2) Amino Acids: Increased plasma amino acids stimulate secretion
      (3) Sympathetic nervous system: Enhances glucagon secretion
      (4) Somatostatin: Decreases glucagon secretion
      (5) Insulin: Decreases glucagon secretion

d. Somatostatin
    i. First identified as growth hormone–inhibiting hormone (from the hypothalamus)
      (1) Produced by the pancreas delta cells and other GI locations
      (2) Stimuli for release are the same as insulin (increased plasma glucose and increased amino acids)
    ii. Actions
      (1) Decreases gut motility
      (2) Decreases secretions by the gut
      (3) Inhibits glucagon and insulin secretion
      (4) Acts as a brake on the GI tract; slows things down (so that everything does not all go through at once and the nutrients can be properly absorbed).

e. Pancreatic Polypeptide
    i. Release: Stimulus is the ingestion of a protein-rich meal, hypoglycemia, or exercise
    ii. Actions: Inhibits the gallbladder and the exocrine pancreas

f. Diabetes Mellitus: Most common endocrinological disorder. Diabetes is the second leading cause of blindness (glaucoma is the leading cause) and the third leading cause of death. Diabetes mellitus = either no insulin or the body does not respond to its own insulin.
    i. Types: Two categories
    (1) Type 1 (Formerly called IDDM)
      a. 10%–20% of diabetics
      b. Progresses rapidly
      c. Usually a child gets a viral disease and type I develops because the T-cells are activated from the original invasion and inappropriately destroy the β-cells of the pancreas.
      d. Some genetic influence (identical twin studies—50% will get it)
      e. *Not* obesity related
      f. Used to be called "juvenile onset"

g. Insulin level = 0

h. Ketosis occurs because they are "starving in a sea of glucose"

(2) **Type 2 (Formerly called NIDDM)**

a. 80%–90% of diabetics

b. Used to be called "adult onset"

c. Symptoms progress slowly (over months or years)

d. Insulin receptors become less sensitive than they were

e. Typically is obesity related

f. Significant genetic component (identical twin studies—100% will get it)

g. Insulin levels may be increased, decreased, or normal

h. Typically there is no significant ketosis (or not as significant as with type 1)

i. **More and more kids are developing this because they are much more sedentary than in the past (TV, video games, eating more, etc.).**

ii. **Symptoms:**

(1) Hyperglycemia: People with diabetes have plasma concentrations of glucose that are so high that their tubular load exceeds their T_{max}.

(2) Diuresis (polyuria)

(3) Glucosuria

(4) Very thirsty and drinking a lot (polydipsia)

(5) Polyphagia (hungry all of the time but the food and energy are not used)

(6) Ketoacidosis (more so in type I because insulin strongly inhibits hormone-sensitive lipase [HSL])

(7) Hyperlipidemia: Increased plasma lipid levels put them **at risk for atherosclerosis.**

iii. **Long-term Pathology**

(1) **Neuropathies (nerve damage):** Peripheral nerve damage is common if untreated. This is probably due to osmotic damage to Schwann cells (because of the high extracellular glucose levels). This presents as sensory loss in the extremities and ulcerations in the feet and ankles.

(2) **Nephropathies (renal failure):** The capillaries in the glomerulus thicken, interfering with filtration.

(3) **Microangiopathies:** Thickening in the capillary membrane leads to poor circulation in the periphery and necrosis of tissue (may necessitate amputation)

(4) **Macroangiopathies:** Atherosclerosis in the renal artery and hypertension; increased risk of CVD

(5) **Retinopathies:** Affects the microcirculation that leads to the retinas. This impaired blood flow presents with microaneurysms in the eyes that may burst.

(6) **Impotence:** Microcirculation here is damaged. Many times diabetes is diagnosed because of this.

(7) **Other visual problems:** Diabetic patients have an increased risk of cataracts. Part of the lens gets carboxylated.

What causes all of these pathologies?

Many things get glycosylated inappropriately due to the high plasma glucose. Thus, these things thicken and circulation is impaired. Hemoglobin also is partially glycosylated. Thus, we can now easily assess patient compliance.

iv. **Insulin and Glucagon**

(1) Interestingly enough, exposure to endogenous insulin suppresses the glucagon response during fasting. So for type 1 diabetics, the glucagon response is also impaired. Thus, they are at risk for hypoglycemia.

    v. **Treatment:** Type I patients will need an exogenous source of insulin. Type II patients may need insulin and to be made more sensitive to its effects. Exercise and diet are both very effective in helping diabetics cope with their disease.

    vi. **Drugs**

        (1) Sulfonylureas: Useful only in patients with exposure to some endogenous insulin.

            a. **(GLUCOTROL)**
                - This both enhances the synthesis of insulin by the pancreas and increases the density of insulin receptors on the target tissue.

        (2) Biguanides: For type II diabetics

            a. **(GLUCOPHAGE)**
                - This decreases hepatic output (gluconeogenesis) and decreases intestinal absorption of glucose.

        (3) Drugs that increase insulin sensitivity (for type II diabetics)

            a. **(AVANDIA)**

        (4) Drugs which mimic incretins (substances produced by the GI system that decrease blood glucose)

            a. **(BYETTA)**
                i. Decreases gastric emptying
                ii. Decreases hunger
                iii. Increases insulin secretion
                iv. Decreases glucagon secretion

        (5) Insulin: Type I diabetics have to have insulin injected. With recombinant DNA, we can now use human insulin (made from bacteria) instead of animal insulin.

            a. **(NOVOLOG)**
            b. **HUMULIN (human insulin)**
            c. **EXUBERA (human insulin)**
            d. **HUMALOGA (insulin lispro); fast-acting**
            e. **LANTUS (insulin glargine); long-lasting**

SUGGESTED READINGS

1. Guyton, A.C., and Hall, J.E. 2011. Chapter 78 in *Textbook of Medical Physiology*, 12th ed. Baltimore: Saunders.

CHAPTER 16 QUESTIONS

1. _____ The following are known to stimulate insulin secretion
   A. Parasympathetic stimulation   D. All of these
   B. Glucagon   E. Two of these
   C. Increased plasma amino acids

2. _____ Insulin
   A. Makes one hungry
   B. Increases the activity of lipoprotein lipase
   C. Stimulates the activity of glucokinase
   D. All of these
   E. Two of these

3. _____ Pancreatic hormone that decreases GI activity
   A. Secretin   D. Glucagon
   B. Cholecystokinin   E. Gastrin
   C. Somatostatin

4. _____ Insulin
   A. Is produced by the islet cells of the pancreas
   B. Is a protein
   C. Is also referred to as pancreatic polypeptide
   D. All of these
   E. Two of these

For questions 5–8, choose one of the following:
   A. Increases or is greater than
   B. Decreases or is less than
   C. Has no effect or is equal to

5. _____ The effect of insulin on gluconeogenesis

6. _____ The effect of insulin on protein synthesis in skeletal muscle

7. _____ The effect of elevated amino acids on glucagon secretion

CHAPTER 17

Endocrinology V: Thyroid and Parathyroid Glands

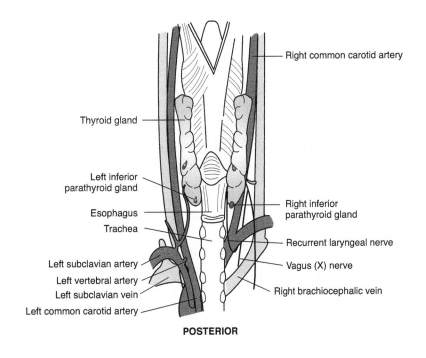

Right common carotid artery

Thyroid gland

Left inferior parathyroid gland

Right inferior parathyroid gland

Esophagus

Trachea

Recurrent laryngeal nerve

Left subclavian artery

Vagus (X) nerve

Left vertebral artery

Left subclavian vein

Right brachiocephalic vein

Left common carotid artery

POSTERIOR

FIGURE 17.1

THYROID
I. **Introduction and Anatomy:** Located immediately below the larynx on the ventral surface of the trachea. This gland has two lobes, one on either side (connected in the middle). This gland is extremely well vascularized. Also note that the parathyroid glands are right below this.

a. **Follicles and Colloid**
  i. The functional unit is the follicle (shown in figure below).
  ii. The epithelial cells are of differing shapes, indicating their varying activities within the gland.
     (1) Squamous (flat) cells = thyroid that is inactive
     (2) Cuboidal cells = thyroid that has intermediate activity
     (3) Columnar (elongated) cells = thyroid that is very active
  iii. **Colloid:** A very viscous, gel-like fluid that stores thyroid hormone (primarily composed of glycoprotein called thyroglobulin). This thyroglobulin contains the various thyroid hormones in various stages of development.
  iv. **Parafollicular cells:** These cells are located between the follicles. These cells (also called "C cells") have a role in secreting calcitonin, which plays an important role in Ca^{++} metabolism.

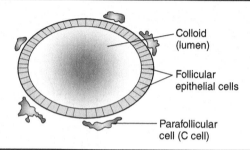

Colloid (lumen)

Follicular epithelial cells

Parafollicular cell (C cell)

FIGURE 17.2

b. Thyroglobulin stores **tetraiodothyronine** (thyroxine or T4) and **triiodothyronine** (T3 or rT3).
  i. T3 is the biologically active form. rT3 (an isomer) has no activity.
  ii. Tyrosine and iodine form the thyroid hormones.
  iii. We will refer to all forms as "thyroxine."

II. Synthesis
a. **Synthesis of Thyroglobulin:** First made at the rough ER, it is then transported to the Golgi apparatus, where it is glycosylated, packaged, and exocytosed into the colloid for storage.
  i. Thyroglobulin is a large protein.
  ii. It contains tyrosine residues, which are used to make thyroid hormones.
b. **Iodide Trapping by the epithelial cells**
  i. Iodide (I^-) is actively transported from the plasma into the epithelial cells against the concentration gradient (can transport against a gradient up to a ratio of 40:1). Thus, you can have a low iodide concentration and still get it into the cell. We can visualize this gland by using **radioactive tracers.**
  ii. **Drugs** can be used to block iodide trapping. Thiocyanates compete for the carrier molecule that transports the iodide and thus reduces T3 and T4 levels.
c. **Oxidation of Iodide:** Iodide is oxidized into iodine (I^- into I).
  i. The enzyme **thyroid peroxidase** (TPO) catalyzes this reaction.
  ii. **Thiouracils (PROPACIL)** can be used to block this activity.

    d.   **Organification:** Iodine attaches to tyrosine residues (part of the thyroglobulin molecule)
        i.   Iodine attaches to tyrosine to form MIT (monoiodothyronine) or DIT (diiodothyronine).
       ii. DIT + MIT = **T3**
           (1) This is only 10% of the hormones secreted.
           (2) Can also form rT3 this way (less than 1%)
      iii. DIT + DIT = **T4**
           (1) This is 90% of the hormones secreted here.
    e.   **Storage:** A significant amount of thyroglobulin can be stored (several months' worth).

III. **Secretion:** Secreted in order to be used by the body
    a.   **Follicular Cells:**
        i.   These bite off (endocytose) a piece of the colloid. It is then digested and released into the plasma.
    b.   **Thyroglobulin** is converted into amino acids, T3, reverse T3, MIT, and DIT:
        i.   T3 and T4 are secreted into the plasma.
       ii. Amino acids, MIT, and DIT are recycled and used again in the formation of thyroglobulins and thyroid hormones.

IV. **Transport & Release to tissues:** Thyroid hormones are quite small and thus must attach to a protein to keep from being filtered.
    a.   **Thyroxine binds globulin.**
    b.   **Thyroxine binds transthyretin.**
    c.   **Thyroxine binds albumin.**
    d.   **Only the free hormone has the biological effect; there is equilibrium between free hormone and bound hormone.**

V. **Biological Activities**
    a.   **T3 and T4 have varying levels of activity.** The thyroid gland produces about 10% T3 and about 90% T4. T3 is 4–5 times as biologically active as T4.
    b.   **Peripherally,** 55% of the T4 is converted to reverse T3 (rT3), which has no biological activity. The remainder of T4 (about 45%) is converted into T3 in the liver.
        i.   There is question as to whether T4 does anything at all.
       ii. T4 may just act as a prohormone.
    c.   **Weight Loss:** Usually people on starvation diets can lose weight quickly, but then after awhile it is harder and harder to lose the weight. Why?
        i.   The body's metabolic rate slows (so it burns fewer calories). How?
       ii. The **enzyme that converts T4 to T3 is inhibited by starvation.** When the food intake slows, it causes the metabolic rate to slow.
    d.   **Estrogens** increase the amount of thyroxine that is bound (with the same amount existing as free hormone).
    e.   **Drugs: DILANTIN** (phenytoin; for seizure disorders) and **salicylates** (e.g., **aspirin**) decrease the binding of the thyroxine so that more exists free in the plasma.

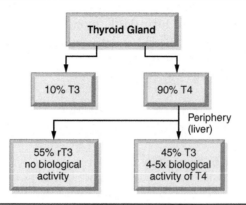

FIGURE 17.3

VI. **Regulation of Thyroid secretion (this is a classic negative feedback system: T4 and T3 levels regulate both TRH secretion and TSH secretion).**
   a. TRH (thyrotropin-releasing hormone) is released from the hypothalamus to the anterior pituitary gland. There it causes the release of **TSH (thyroid-stimulating hormone), which then has a number of effects on the thyroid tissue.**
      i. Iodide trapping: TSH enhances iodide trapping.
      ii. Organification: Enhances the synthesis of T4 and T3.
      iii. Endocytosis of colloid: Enhances the endocytosis of the colloid by the follicular epithelial cells
      iv. Proteolysis of the thyroglobulin: Enhances the digestion of thyroglobulin (to sort out T4 and T3)
      v. Release of T3 and T4: Enhances the release of T3 and T4
      vi. The net effect is that the thyroid gland is more active.

VII. **Actions of T4/T3**
   a. **Maturational/Differentiation effects:**
      i. T3 and T4 have a maturational effect or differential effect.
      ii. **Amphibians:** Developing frogs become tadpoles and then frogs (T3 on a frog will enhance the maturation rate).
      iii. **Humans:** Thyroid hormones are needed for normal maturation.
         (1) If children are deprived of thyroid hormone, they will have delayed bone maturation.
         (2) Hypothyroidism: Child may never experience puberty (or it may be delayed) if thyroid hormones are not present. Thyroid hormones are also needed for normal surfactant production.
   b. **Neurological effects:**
      i. During normal fetal and neonatal development, there is a critical period where the neonate must be exposed to thyroid hormones.
         (1) **The critical period** begins a few months before birth and ends around 2 years of age.
         (2) If children are not exposed to thyroid hormones at this time, there are irreversible damages.
      ii. **Adults that exhibit hypothyroidism:**
         (1) Have prolonged reflex times
         (2) Become lethargic
         (3) Have impediments to neurological function

c. **Growth**
  i. T3 and T4 are necessary for normal growth.
    (1) There is stunted growth when thyroid hormones are deficient.
  ii. GH and IGF decrease if thyroid hormones are low. For hGH and IGFs to work properly, we must have thyroid hormones present!
d. **Metabolic effects**
  i. T3 and T4 are needed to regulate BMR. They increase BMR by increasing mitochondrial surface area and increasing oxidative phosphorylation.
  ii. Hyperthyroidism: Increased BMR
  iii. Hypothyroidism: Decreased BMR
  iv. Involved in stimulating both anabolic and catabolic pathways.
e. **Sympathomimetic effects**
  i. Thyroid hormones increase the beta receptor density and beta receptor affinity for epinephrine and norepinephrine.
f. **Effects on reproduction**
  i. T3 and T4 are needed for normal ovarian cycling or follicular development.
  ii. Spermatogenesis is also thyroid dependent.
  iii. Various reproductive problems when the thyroid is altered (viewed more as a **permissive role in regards to reproduction**)

VIII. **Pathologies**
  a. **Hypothyroidism**
    i. **Causes**
      (1) Primary hypothyroidism: A pathology of the gland itself. Main cause is Hashimoto's thyroiditis, an autoimmune disorder.
      (2) Secondary hypothyroidism: TRH or TSH (made by the anterior pituitary) related deficiency
      (3) Nutritional hypothyroidism: Typically due to decreased iodine in the diet
    ii. **Symptoms: Child vs. Adult**
      (1) Children: Experience more symptoms; more disastrous
        a. Respiratory distress syndrome (RDS)
        b. Impaired bone growth
        c. Mental retardation
        d. Stunted growth
        e. Delayed or absent sexual maturity
        f. **Cretinism:** Collective term for disease of hypothyroidism in children, as indicated by symptoms above
      (2) Adults: Fewer symptoms
        a. Decreased BMR
        b. Poor resistance to cold
        c. An increase in weight
        d. Slow thinking
        e. Thin or brittle hair
        f. Decreased information processing
        g. Weakness, fatigue, lethargy
        h. **Myxedema:** an unusual edema that is caused by the accumulation of mucopolysaccharides in the interstitial spaces (these exert an osmotic pressure)

    iii. **Treatment of Hypothyroidism**
        (1) Give patient thyroxine preparation
        (2) Drugs: (LEVOXYL) and **(SYNTHROID)** are among the top five drugs prescribed each year.
  b. **Hyperthyroidism** or "thyrotoxicosis," as the effect of it is known.
    i. **Causes:** The most common cause is **Graves' disease, an autoimmune disorder:**
        (1) Body erroneously produces TSI (thyroid-stimulating immunoglobulin), which acts just like TSH, but there is no negative feedback system to shut it off.
    ii. **Symptoms**
        (1) Increased BMR
        (2) Intolerant of heat
        (3) Increased appetite
        (4) Increased HR (arrhythmias or palpations)
        (5) Nervous or irritable
        (6) Exophthalmos: Eyes bug out
    iii. **Treatment**
        (1) Surgically: Remove all or part of the thyroid.
        (2) Chemically: Use iodine 131 (^{131}I), a radioactive isotope that is picked up by the thyroid and stored. It is a gamma ray emitter, and thus it destroys part of the thyroid tissue.
        (3) Thiouracil (**PROPACIL**): Inhibits thyroid hormone synthesis.
  c. **Goiter**
    i. An enlarged thyroid gland.
    ii. Could be due to **hypothyroidism or hyperthyroidism.**
    If one effect that TSH has on the thyroid tissue is that it increases growth of the tissue, explain how both hyposecretion of T3 and T4 and hypersecretion of T3 and T4 in Graves' disease could both lead to goiters.

Parathyroid Hormone (Calcitonin, Calcium, and Phosphate)
I. **Absorption and Excretion of Calcium and Phosphate**
  a. **Calcium (Ca^{++})**
    i. Milk products are the major source of calcium.
        (1) Calcium is naturally occurring in products such as milk and cheese.
    ii. Problems: Calcium is poorly absorbed in the gut.
        (1) Only 50% of the calcium consumed is absorbed in the gut.
    iii. Calcium is also excreted in the feces and in the urine (like Na^+).
        (1) A great deal of calcium is reabsorbed in the proximal convoluted tubule (about 98%), which is under no regulation.
        (2) The last 2%–3% can be reabsorbed in the distal convoluted tubules, but it depends on the needs of the body.
  b. **Phosphate**
    i. Phosphate is absorbed well in the gut, unless there is excess calcium present.
    ii. Phosphate is excreted in the urine in a manner similar to glucose.
        (1) As long as the T_{load} is low, all of the phosphate that is filtered gets reabsorbed.
        (2) If the T_{load} is greater than the T_{max}, then the excess phosphate gets excreted (excess phosphate is lost in the urine).

II. Role of Vitamin D

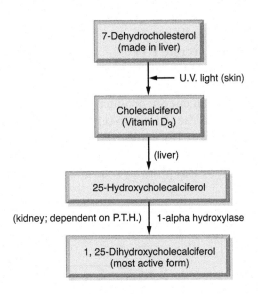

FIGURE 17.4

a. There are several compounds in the Vitamin D family (of which cholecalciferol is the most important).
   i. Vitamin D3 (cholecalciferol) can be obtained in the diet or from *de novo* synthesis in the liver (many foods are fortified with Vitamin D).
   ii. Vitamin D3 can be synthesized from a cholesterol precursor
      (1) 7-dehydrocholesterol in combination with UV light (from the skin) can make Vitamin D3.
      (2) Vitamin D3 then must travel to the liver to become biologically active.
      (3) In the liver Vitamin D3 becomes 25-hydroxycholecalciferol.
      (4) 25-hydroxycholecalciferol then goes to the kidney and in the presence of the enzyme 1-alpha hydroxylase becomes 1,25 dihydroxycholecalciferol, the most active form of Vitamin D. This last conversion in the kidney is dependent on parathyroid hormone (PTH), meaning it is dependent on the needs of the body.
b. 1, 25-dihydroxycholecalciferol is the most active form of Vitamin D.
   i. Its activity is most predominant in the gut.
   ii. In the gut, this form of Vitamin D induces the formation of calbindin, which promotes the absorption of both Ca^{++} and PO_4^{3-}.
   iii. Also in the kidney in synergism with PTH, Vitamin D enhances calcium reabsorption.
   If someone gets plenty of sunlight, do they need Vitamin D in their diet?

What if they live in New York and it is the middle of the winter?

III. **Parathyroid Hormone (PTH):** This is a small protein (84 amino acids) that is made in the parathyroid gland. There are four parathyroid glands (behind the thyroid gland), which are very small. There is an intimate relationship between the parathyroid and the thyroid gland. If we take the thyroid gland out, we can easily damage the parathyroid glands because they are so small. The glands are made up of two pairs of glands (four total):

  a. **Morphology of Gland: Two cells types**
    i. **Chief Cells:** Produce and secrete PTH
    ii. **Oxyphil Cells:** Their function is unclear, but they are not involved in parathyroid hormone production.
      (1) They may be involved in energy production or storage.
      (2) They are not present in the parathyroid gland until puberty.
  b. **Parathyroid Hormone Actions:** If we inject PTH, we see an increase in plasma concentrations of Ca^{++} and a concurrent decrease in plasma concentrations of phosphate. PTH increases Ca^{++} levels and decreases phosphate levels.

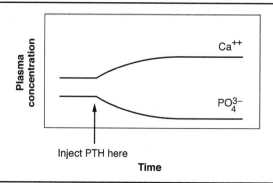

FIGURE 17.5

  i. **Increases plasma calcium:**
    (1) **Bone is constantly being remodeled.** Osteoclasts break bone down and release Ca^{++} while osteoblasts lay down new bone (using Ca^{++}). If we increase the activity of the osteoclasts (an effect of PTH), then we break bone down and increase the plasma concentration of Ca^{++}.
    (2) **PTH also enhances the reabsorption of Ca^{++}** in the DCT (distal convoluted tubules), which decreases the renal clearance of Ca^{++}, and thus more Ca^{++} is put back in the plasma.
    (3) **PTH also enhances the synthesis of Vitamin D, which** enhances the absorption of Ca^{++} at the gut level. This increases the plasma concentration of Ca^{++}.
    ii. **Decreases plasma phosphate:** PTH decreases phosphate reabsorption in the proximal convoluted tubules by altering the T_{max}. This is not as efficient as Ca^{++} reabsorption. The renal clearance of phosphate increases.
  c. **Regulation**
    i. Decreased plasma calcium is the strongest PTH stimulus. This stimulates the parathyroid glands to increase PTH production.
    ii. Ca^{++} levels are monitored very carefully. Maintenance of plasma Ca^{++} levels is very important (more important than maintaining bone Ca^{++}).
    iii. Increased Ca^{++} levels suppress PTH secretion.

IV. **Calcitonin (Thyrocalcitonin)**
   a. **History**
      i. Produced by the parafollicular cells of the thyroid.
      ii. Calcitonin is a polypeptide made up of only 32 amino acids.
      iii. There is not a lot of variety (transferable) between species:
         (1) Other species' calcitonin is more effective in humans than our own calcitonin.
         (2) Calcitonin activity is particularly important in animals that produce eggs, or live where there is low Ca^{++} (fresh water fish).
         (3) **Salmon calcitonin is 20 times more potent than humans' (and thus we must give a nonphysiological dose of human calcitonin to see an effect in humans).**
   b. **Action:** Opposite from parathyroid hormone
      i. Lowers the Ca^{++} plasma levels by **inhibiting osteoclastic activity**
      ii. Its role in humans is speculative. **There are no definitive complications from an excess or a deficiency in calcitonin.**
      iii. Calcitonin is mostly used for osteoporosis therapy
         (1) Usually this is in the form of a nasal spray because it is biodegradable in the stomach or SI.
         (2) MIACALCIN (salmon calcitonin) is used.
      iv. Increasing the plasma concentration of Ca^{++} stimulates calcitonin secretions.
      v. Some GI hormones stimulate calcitonin secretions (due to the presence of food in the gut).

V. **Pathologies**
   a. **Hypocalcemia**
      i. Could be due to damage during thyroid surgery
      ii. **Symptoms**
         (1) Progressive hyperexcitability
         (2) Tingly fingers
         (3) Cardiac arrhythmias
      iii. **Tests for hypocalcemia**
         (1) **Chvostec test:** Tap at an angle of the jaw (trigeminal nerve). If there is too little Ca^{++}, then twitches will result.
         (2) **Trousseau test:** Inflate a BP cuff and watch for spasms and contractions in the hand (nerves hyperexcitable).
   b. **Hypercalcemia**
      i. Could be due to a parathyroid tumor
      ii. **Symptoms:** Depressed CNS and muscle function (hypoexcitable)
   c. **Rickets** (not seen much these days)
      This used to be fairly common. The classical case is in children in parts of the world where there are long, prolonged winters. Because the children do not get enough sunlight, they cannot synthesize enough Vitamin D3 *de novo* in the liver. These children can easily become Vitamin D deficient, and thus, the Ca^{++} from their diet is suddenly severely impaired (remember, we need Vitamin D to absorb Ca^{++} in the gut). These children must then secrete PTH and consequently take Ca^{++} from the bones. As a result, these children develop low bone Ca^{++}, and their legs begin to "bow out"/ bend.
   d. **Osteomalacia** (adult Rickets—a different etiology)
      i. As we age, there can be trouble absorbing fats.
      ii. Vitamin D is fat soluble, so Vitamin D is not absorbed as well in the diet.
      iii. There is low plasma Ca^{++} as a result of the Vitamin D deficiency and subsequent PTH secretion.
      iv. This disease does not progress as far as rickets does.

e. **Osteoporosis**
    i.   Bone density decreases due to lack of estrogens (usually in postmenopausal women).
    ii. Treatment: Synthetic estrogens to encourage osteoblastic activity.

SUGGESTED READINGS

1. Guyton, A.C., and Hall, J.E. 2011. Chapters 76 and 79 in *Textbook of Medical Physiology*, 12th ed. Baltimore: Saunders.
2. Koeppen, B.M. and Stanton, B.A. 2008. Chapters 35 and 41 in *Berne & Levy Physiology*, 6th ed. St. Louis: Mosby.

CHAPTER 18

Endocrinology VI: Adrenal Physiology

I. **Introduction:** The adrenal glands sit on top of the kidneys (suprarenal) and are very small. They are composed of two regions.
   a. **Medulla:** The innermost part of the gland
      i. It makes up 20% of the gland.
      ii. It secretes epinephrine and norepinephrine (predominately epinephrine).
      iii. The medulla is derived from the ectoderm.
   b. **Cortex:** The outermost region of the gland. The cortex comprises the majority of the gland (80%). It produces and secretes corticosteroids (steroidal in nature and from the cortex). The cortex is derived from the mesoderm. It is separated into three regions.
      i. **Zona glomerulosa:** The outermost part of the cortex
         (1) Produces the mineralocorticoids (a group of hormones)
         (2) **Aldosterone** is the most predominant mineralocorticoid.
      ii. **Zona fasciculata:** The middle region of the cortex, and the largest
         (1) Produces a group of hormones, the glucocorticoids
         (2) The predominant hormone is **cortisol.**
         (3) Also can produce some androgens
      iii. **Zona reticularis:** The innermost region
         (1) Major product of this layer is the androgen **DHEA**
         (2) Also can produce some cortisol
      iv. Under normal conditions, estrogen and androgens are not sufficiently potent or in high enough quantity to cause a masculine or feminine effect.
         (1) **The only androgen that has any effect is DHEA** (dehydroepiandrosterone). DHEA is responsible for the distribution of body hair on females to the axillary and pubic regions.
   c. **Pathologies associated with adrenal androgen hypersecretion:** The abnormal secretion of androgens from the adrenal gland (**adrenogenital syndrome**). With this there are excess androgens

and estrogens being secreted from the adrenal glands. This varies with age and gender. **We will focus on an excess of androgens in females.**

    i. **In adult females:** Causes masculinization; they may experience a deepening of the voice and become hairy ("hirsutism") around the face.

    ii. **In the female embryo:** Effects are more profound

        (1) **Female pseudohermaphroditism:** In this condition, a female fetus develops male external genitalia (but has ovaries).

    iii. **In adult males:** There is no apparent effect, because they are already exposed to androgens from their testicular androgens.

    iv. **In young boys:** "**precocious pseudopuberty**" results (puberty at an early age)

        (1) Occurs in 6- to 7-year-old boys

        (2) Deepening of the voice and pubic hair

II. The mineralocorticoids (Aldosterone)

a. Functions

    i. **Primary Function:** Promotes the reabsorption of Na^+ and secretion of K^+ at the distal portions of the nephron (DCT and collecting duct)

        (1) 3 Na^+ reabsorbed for every 2 K^+ secreted

b. Regulation of Secretion

    i. **Renin-Angiotensin:** Renin converts angiotensinogen to angiotensin I and then ACE converts this into angiotensin II. Angiotensin II is the stimulus.

        (1) **Angiotensin II causes thirst and the secretion of aldosterone** (increases synthesis and release).

    ii. K^+: **High levels of plasma K^+ induce aldosterone secretion.**

        (1) Has a direct effect on the adrenal gland itself

    iii. **ACTH (adrenocorticotropic hormone):** This has no direct effect on inducing aldosterone secretion.

        (1) The zona glomerulosa does have ACTH receptors, but it is not an important regulator.

        (2) ACTH is necessary for normal growth and maintenance of the gland, but it has no effect on regulating the secretion of aldosterone.

    iv. **Increased plasma Na^+:** This very slightly decreases aldosterone secretion.

c. Pathologies

    i. **Hypersecretion** (from the adrenal gland itself)

        (1) **Primary hyperaldosteronism (Conn's Syndrome):** This is usually due to an aldosterone secreting tumor on the gland itself.

        (2) **Secondary hyperaldosteronism:** This is not due to the gland itself, but rather is generally due to an error in the renin-angiotensin system. Typically this is due to atherosclerosis in the renal arteries, so the GFR decreases, and thus the macula densa sense this. They then signal the JG cells to release renin (to increase GFR). The eventual formation of angiotensin II causes the secretion of aldosterone. This is a dangerous feed-forward loop.

    ii. **Hyposecretion:** A patient can have hypersecretion in only one zone, but typically hyposecretion is usually all zones (will be discussed with glucocorticoids).

III. The Glucocorticoids (Cortisol): There are several glucocorticoids, but in humans the most important is cortisol.

a. Metabolic Effects: Cortisol converts proteins and fats to glucose (gluconeogenic) and increases plasma glucose levels.

i. **CHO metabolism**
   (1) Increases gluconeogenesis
   (2) Decreases glucose utilization by cell (anti-insulin effect), especially by muscle and adipose tissue
ii. **Protein metabolism**
   (1) Proteins are broken into amino acids and converted to glucose
   (2) Lean body mass decreases (because of the above)
iii. **Fat metabolism**
   (1) Increases lipolysis
   (2) Enhances catecholamines' effect on lipolysis
   (3) "Buffalo hump": unusual fat distribution; an accumulation of fat between the scapula on patient's back
iv. **Hunger**
   (1) Increases appetite/hunger

b. **Other Effects**
   i. **Fetal**
      (1) Aids maturation of several structures (e.g., lungs; needed for surfactant production).
      (2) Involved in maturation of GI enzymes (which develop late in pregnancy)
   ii. **Adults**
      (1) **Stress**
         a. **Hans Selye (1900s):** Experimented by stressing animals and noted that stressed animals had increased glucocorticoid levels. This was part of the general adaptation syndrome. He then performed this over a wide variety of stimuli and found that animals that live in polluted or noisy environments had increased glucocorticoids as well. Why? Perhaps because glucocorticoids are strongly gluconeogenic.
      (2) **Anti-inflammatory**
         a. In nonphysiological doses, glucocorticoids are anti-inflammatory. During tissue damage, there is increased blood flow to the area and capillaries become more permeable. Prostaglandins and leukotrienes come to the site (responsible for inflammation). During tissue damage, phospholipase A2 activity increases the release of arachadonic acid to aid in the synthesis of prostaglandins and leukotrienes (these are thought to initiate the inflammatory response). Glucocorticoids enhance the production of **macrocortin**, which inhibits phospholipase A2 and thus the inflammatory response.
      (3) **Immune response**
         a. Also in nonphysiological doses, cortisol decreases the number of T-lymphocytes and thus suppresses the immune system. Cortisol is for this reason often used in patients who have had organ transplants. One can also treat allergy-type responses by this same mechanism.
      (4) **Vasoconstriction**
         a. In order for the catecholamines epinephrine and norepinephrine to work properly in the vasoconstrictive response, glucocorticoids have to be present.
      (5) **Erythropoiesis**
         a. Stimulates erythropoietin (increases hematocrit)
      (6) **Bone**
         a. Increases bone resorption
      (7) **REM sleep**
         a. Decreases REM sleep

c. **Regulation**
   i. A classic negative feedback system
   ii. **Decreased glucocorticoid levels** cause hypothalamus to secrete CRH (corticotrophin-releasing hormone).
      (1) CRH travels down the hypophyseal portal vein to the anterior pituitary, causing it to release ACTH.
      (2) CRH and low glucocorticoid levels cause the anterior pituitary to release ACTH.
      (3) ACTH stimulates glucocorticoid production at the adrenal cortex.
   iii. **Stress and hypoglycemia can also trigger the release of CRH.**

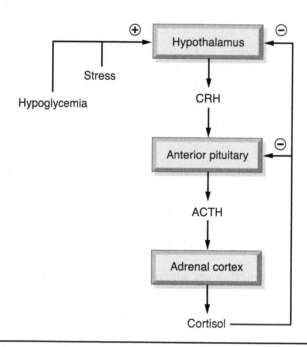

FIGURE 18.1

d. **Pathologies**
   i. **Hypersecretion (Cushing's Syndrome)**
      (1) 1 out of 1,000 hospital-admitted patients suffer from this.
      (2) More common in females
      (3) Caused by too much exogenous cortisol (taken for organ transplantation), too much ACTH, an adrenal tumor, or ACTH-secreting tumor
      (4) Symptoms: The breakdown of proteins (proteolysis) and unequal fat distribution (e.g., "moon facies" or "buffalo hump")
   ii. **Hyposecretion**
      (1) Primary Adrenal Cortical Insufficiency (Addison's Disease):
         a. Due to autoimmune destruction of the gland. Adrenal cortex gets destroyed. The most acute effect is the loss of aldosterone (mineralocorticoid problem), although both are affected.
      (2) Secondary Adrenal Cortical Insufficiency:
         a. There is too little ACTH (not a problem with the gland itself). These patients respond to stress poorly and may be hypoglycemic.

In which disease might a patient present with a darker skin color? Why?

IV. Prostaglandins
  a. **Introduction**
    i. **Eicosanoids** are a class of compounds that **arachidonic acids belong to.**
      (1) Compounds are formed from polyunsaturated fatty acids (18, 20, 22 carbons).
      (2) The most important eicosanoid is arachidonic acid, which is used to synthesize prostaglandins, leukotrienes, prostacyclins, and thromboxanes.

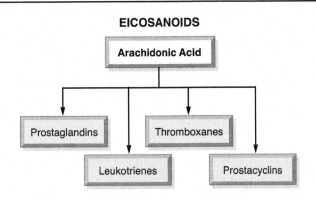

FIGURE 18.2

    ii. **Prostaglandins** are not considered to be hormones because they are not synthesized in only one place and they are used locally.
      (1) They function either as paracrines or as autocrines.
      (2) Prostaglandins are locally active and have a short-term effect.
    iii. **Prostaglandin Nomenclature**
      (1) First identified in the semen and thought to be from the prostate gland, so the name stuck

  b. **Classification:** Belong to one of three groups
    i. PGA
    ii. PGE
    iii. PGF
      (1) Further identified by the number of double bonds in the side chains.
      (2) Further nomenclature with a Greek letter, which indicates which optical isomer it is.
        a. Example: PGE2α (PGE, with 2 double bonds, α isomer)

  c. **Functions:** It is hard to tell what they do not do. They have a number of effects in different systems
    i. **Reproductive System**
      (1) Sperm transport
      (2) Ovulation

        (3) Parturition

        (4) Menstruation

   ii. Respiratory System

        (1) Can act as bronchiole constrictors or bronchiole dilators

   iii. Nervous System

        (1) Mechanism for thermal regulation in hypothalamus

        (2) Influence the release of neurotransmitters

   iv. Immune System

        (1) Increase the inflammation response

   v. Many other functions

V. Drugs: There are a number of drugs that work to inhibit prostaglandin synthesis.

   a. Aspirin, ibuprofen

      i. Many nonsteroidal anti-inflammatory drugs (NSAIDs) work as prostaglandin inhibitors.

      ii. Antipyretics also involve prostaglandin inhibition (Aspirin).

      iii. Therapeutic for dysmenorrhea or menstrual cramps

      iv. These work best if taken prophylactically.

SUGGESTED READINGS

1. Guyton, A.C., and Hall, J.E. 2011. Chapter 77 in *Textbook of Medical Physiology*, 12th ed. Baltimore: Saunders.

2. Koeppen, B.M. and Stanton, B.A. 2008. Chapter 42 in *Berne & Levy Physiology*, 6th ed. St. Louis: Mosby.

Name_____ Date _____

CHAPTER 17 AND 18 QUESTIONS

1. _____ An autoimmune disease involving the thyroid
   A. Graves' disease
   B. Myxedema
   C. Cretinism
   D. Addison's disease
   E. None of these

2. _____ Most active form of Vitamin D
   A. 25-hydroxycholecalciferol
   B. Cholecalciferol
   C. 5, 25 dihydroxycholecalciferol
   D. 7-dehydrocholesterol
   E. 1, 25 dihydroxycholecalciferol

3. _____ Secreted by the thyroid gland
   A. T-4
   B. DIT
   C. Calcitonin
   D. All of these
   E. Two of these

4. _____ Specific part of the adrenal gland that produces cortisol-like compounds
   A. Zona cortex
   B. Zona glomerulosa
   C. Zona reticularis
   D. Zona fasiculata
   E. C and D

For questions 5–11, choose one of the following:
   A. Increases or is greater than
   B. Decreases or is less than
   C. Has no effect or is equal to

5. _____ The effect of TSH on iodine trapping

6. _____ The potency of T-3 as compared to the potency of T-4

7. _____ The effect of SYNTHYROID on the metabolic rate

8. _____ The effect of PTH on the renal clearance of calcium

9. _____ The effect of calcitonin on osteoclastic activity

10. _____ The effect of cortisol on gluconeogenesis

11. _____ The effect of ACTH on aldosterone secretion

CHAPTER 19

Digestion

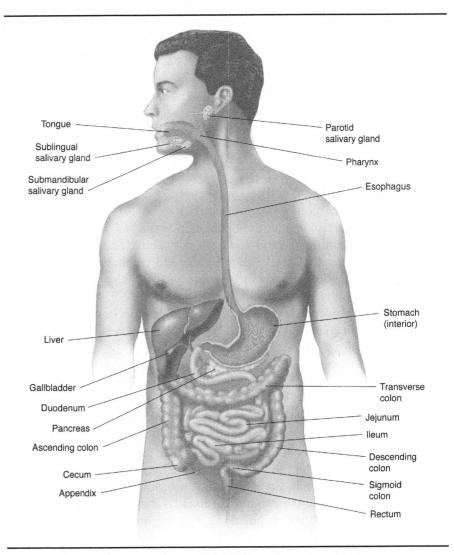

Tongue

Sublingual
salivary gland

Submandibular
salivary gland

Parotid
salivary gland

Pharynx

Esophagus

Stomach
(interior)

Liver

Gallbladder

Duodenum

Pancreas

Ascending colon

Cecum

Appendix

Transverse
colon

Jejunum

Ileum

Descending
colon

Sigmoid
colon

Rectum

FIGURE 19.1

I. **Introduction**
   a. **Anatomy**
      i. GI Tract (alimentary canal): This is a tube from the mouth to the anus that is about 30 feet long. The lumen could technically be considered "outside the body." **Organs of the GI tract:**
         (1) Mouth, pharynx, esophagus, stomach, SI (duodenum, jejunum, ileum), large intestines (ascending colon, transverse colon, descending colon), rectum, anus
      ii. **Accessory Organs:** Salivary glands, liver, gallbladder, pancreas, and many other smaller glands (mucus secreting, etc.)
   b. **Functions**
      i. Food ingestion
      ii. Food movement: From the mouth to the anus **at the appropriate rate** so that digestion can occur. There are mechanisms in place to regulate this rate.
      iii. Mechanical preparation of food: Chewing
      iv. Chemical digestion of food
         (1) Polysaccharides are broken down into monosaccharides so that they can be absorbed
         (2) Proteins are broken down into amino acids (and some small peptides)
         (3) Fats are broken down into monoglycerides and free fatty acids so that they can be absorbed
      v. Absorption of food
      vi. Elimination of food wastes
   c. **Microanatomy:** Layers of the GI tract (these are present from the esophagus to the anus). The similar anatomical structures are continuous throughout the GI tract. There are four layers, or tunicae, starting from the inside (lumen) and going outward.
      i. **Tunica Mucosa:** Has several sublayers
         (1) **Epithelial cells:** This layer serves as a protective surface (in contact with the stuff in the lumen). In some places it is modified for absorption or secretion. This layer also may contain endocrine and exocrine cells.
         (2) **Lamina propria:** Contains connective tissue, blood vessels, and nerve fibers. This layer contains GALT (Gut-Associated Lymphatic Tissue), which is paramount in protecting the body against foreign and harmful substances. The GALT has as much lymph tissue as the whole rest of the body.
         (3) **Muscularis mucosa:** A layer of smooth muscle cells
      ii. **Tunica submucosa:** This contains a lot of connective tissue, blood vessels, and lymphatic tissue.
         (1) Contains a nerve network, the **submucosal plexus (also known as Meissner's plexus). This plexus is responsible for blood flow to the GI tract and secretions of the GI tract.**
      iii. **Tunica muscularis:** A double layer in most of the GI tract; houses two types of muscle
         (1) **Inner circular muscle**
         (2) **Outermost longitudinal muscle**
         (3) In most of the digestive system, these two muscle layers are present (except at the extreme ends, the mouth and anus, where there is skeletal muscle).
         (4) Between the two muscle layers is the **myenteric plexus (also known as Auerbach's plexus),** which functions as an integrating circuit responsible for mechanical movement in the GI tract.
      iv. **Tunica serosa:** The outer layer, made up of connective tissue
         (1) In many places this is continuous with the mesentery (which supports the organs in the abdominal cavity).

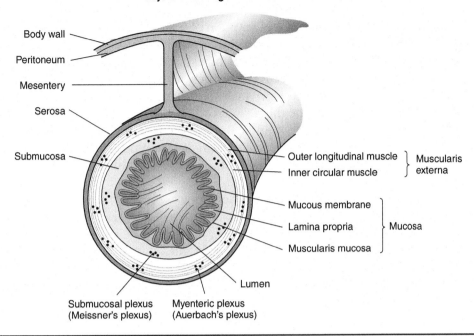

Layers of the Digestive-Tract Wall

Body wall
Peritoneum
Mesentery
Serosa
Submucosa
Outer longitudinal muscle } Muscularis externa
Inner circular muscle
Mucous membrane
Lamina propria } Mucosa
Muscularis mucosa
Lumen
Submucosal plexus (Meissner's plexus)
Myenteric plexus (Auerbach's plexus)

FIGURE 19.2

d. **Gastrointestinal regulation:** Food is digested and moved through the lumen at the appropriate rate; secretions are released at the right time. There is an enormous amount of regulation.
   i. **Automaticity of smooth muscle:** Some muscle cells have automaticity (their own rate or rhythm), meaning they have spontaneous activity (no resting membrane potential; similar to the cardiac muscle).
      (1) The cells most responsible for this phenomenon are modified muscle cells: **the interstitial cells of Cajal (pacemaker cells).**
      (2) The inherent rhythmicity causes slow oscillations; resting membrane potential (RMP) is questionable, but the cells are not at rest.
      (3) This may not be sufficient to cause an action potential, but it at least causes an upward migration of the RMP.
      (4) In general: When food is *not* present, oscillation occurs. When food is present, then neuronal mechanisms cause oscillations.
   ii. **Intrinsic nerve plexus (enteric nerve plexus):** Within the gut there is a network of interconnecting nerves and neurons that are as numerous as within the spinal cord (very complex).
      (1) **Myenteric plexus** has a predominant role in regulating muscle movement in the gut.
      (2) **Submucosal plexus** is responsible for blood flow and secretions of the gut.
      (3) The intrinsic nerve plexus runs from the esophagus to the anus. The GI system is the only organ system with its own private nervous system. Some have called it the "gut brain."
   iii. **Extrinsic nerves:** Neuronal innervation from the ANS
      (1) Motility and secretions can be altered via the ANS.
      (2) **Parasympathetic control:** Increases motility and secretions
      (3) **Sympathetic control:** Decreases motility and secretions
   iv. **Gastro-intestinal hormones:** An endocrine phenomenon; an arsenal of GI hormones are released in response to food in the gut.

II. **Mouth:** Food encounters the teeth, which are designed for chewing; the mechanical digestion of food
   a. **Chewing:** The mechanical digestion. Dentition of teeth depends on the food that an animal eats (herbivores = more dentition; carnivores = less dentition). This process increases the surface area exposed to digestive enzymes by breaking food apart. Humans are omnivorous, and thus a mixture of both. Dogs seem to not chew when they eat. This does not cause a problem because all of the digestive enzymes needed are present in the pancreatic digestive juices. Humans could do the same thing (swallow without chewing), except for such things as vegetables, which may have an indigestible β-glucose coating (we cannot digest β-glucose).
   b. **Saliva:** We produce about 1-2 liters of saliva each day from three pairs of glands.
      i. **Glands:** Parotid (most proliferate), submandibular, and sublingual (least proliferate)
      ii. **Functions**
         (1) **Digestion with enzymes** (mainly the α-amylase ptyalin). Ptyalin breaks down long chain polysaccharides into shorter chains (even disaccharides).
         (2) **Lubricates food:** Mucin combines with water to form mucus, a slimy, slippery lubricant that protects the epithelial layer of the gut.
         (3) **Solvent:** This allows us to taste food.
         (4) **Oral hygiene:** Lysozymes help destroy unwanted pathogens. Immunoglobulins are also present in saliva.

      iii. **Pathologies**
         (1) **Xerostomia:** Insufficient salivary secretion (dry mouth)

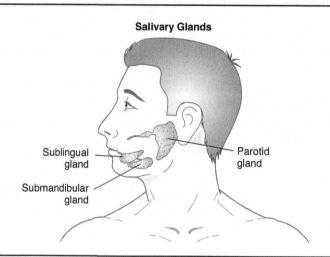

Salivary Glands

Sublingual gland

Submandibular gland

Parotid gland

FIGURE 19.3

III. **Esophagus**
   a. **Swallowing:** Once we swallow food, the digestion in the mouth is done. Swallowing is initiated voluntarily, but then the rest of the trip is under involuntary control (regulated at medullary centers). **Three phases:**
      i. **Oral phase**—Food propelled by the tongue toward the pharynx (voluntary)
      ii. **Pharyngeal phase**—Propulsion of food from the pharynx into the esophagus. The larynx is closed via the epiglottic cartilage.

iii. **Esophageal phase:** From the esophagus to the stomach
   (1) At the esophagus, there is a primary peristaltic wave that propels the food down into the stomach. We also may have a secondary peristaltic wave if the primary wave is not sufficient.
   (2) The esophagus is mostly in the thoracic cavity. The diaphragm separates the thoracic and abdominal cavities. Thus, the esophagus must penetrate through this.
   (3) The esophageal hiatus is the opening through which the esophagus passes from the thoracic cavity to the abdominal cavity.
   (4) Usually this hiatus is a tight fit. If a small tear takes place here, we have a hiatal hernia (and may have problems with digestion and the like).
b. **Pharyngoesophageal sphincter:** Most of the esophagus is exposed to subatmospheric pressure of the thoracic cavity, and thus we must close off the esophagus so as to not aspirate air into this system upon inspiration. Thus it is important that the pharyngoesophageal sphincter remain closed.
c. **Gastroesophageal sphincter:** This is normally kept closed so that the contents of the stomach do not migrate into the esophagus.
   i. Sometimes it does not close properly and we get **gastric reflux (heart burn);** acid chyme and/or food backs into the esophagus.
   ii. When we swallow there are peristaltic **waves that open the gastroesophageal** sphincter.
   iii. **Achalasia:** Condition in which this sphincter does not open properly. With this, food accumulates in the esophagus and the patient may aspirate some food into the respiratory circuit when inhaling.

IV. Stomach

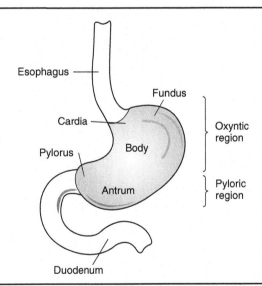

FIGURE 19.4

a. **Anatomy**
   i. **Macroanatomy:** The stomach is between the esophagus and the duodenum. It is divided into three main areas:
   (1) Fundus (upper portion), body (middle), and antrum (lower portion)
   (2) The same basic 4 tunica layers

(3) Mucosa and the submucosa are fused together so that folds are formed, called rugae (this creates a lot more surface area).

(4) Each day the stomach secretes 2 liters of gastric juice.

(5) Gastric juice is secreted by cells that are associated with the gastric mucosa.

ii. Microanatomy: The **gastric mucosa** is divided into two distinct areas.

(1) Oxyntic mucosa: Lines the body and the fundus. The gland-like cells located here have gastric pits associated with them. These gastric pits have three cell types.

    a. Mucous neck cells: These secrete a thin, watery mucus. These cells are highly mitotic and are the progenitors of all other cell types located here. We resurface our stomach with new surface mucous cells about every 3 days!

    b. Chief cells: These produce **pepsinogen.**

    c. Parietal cells: These produce **HCl** and **intrinsic factor.**

    d. Between the pits are the **surface mucous cells.** These produce a thick, viscous mucus. This provides a mucous coating for the inside of the stomach.

(2) Pyloric gland area: The surface in the lower stomach (antrum)

    a. These cells secrete primarily mucus, a little pepsinogen, **but no HCl.** This area also produces **gastrin** and **somatostatin.**

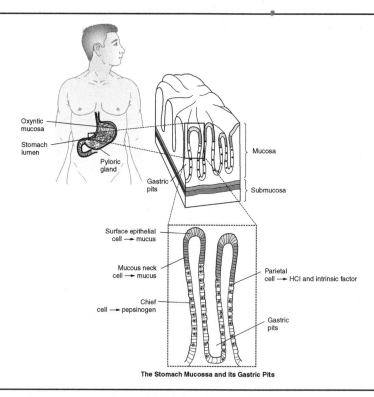

The Stomach Mucossa and its Gastric Pits

FIGURE 19.5

iii. Digestion

(1) Mechanical activity

    a. Storage: The stomach can change its volume to 10 times its resting volume. Food can be stored temporarily in order for it to be just "trickled" into the SI.

    b. Mixing: Mixes food with gastric juice

    c. Movement of food into the duodenum at the appropriate rate

(2) Chemical digestion
    a. Mucus: Provides protection for the lining of the stomach.
    b. Gastritis: People suffering from this should not take anti-inflammatory drugs because they limit prostaglandins (GI protection at this level is prostaglandin dependent). Mucus provides a barrier to keep the other secretions away from the wall of the stomach.
    c. HCl: Produced by the parietal cells. **HCl activates pepsinogen into pepsin,** which can then begin breaking down protein. The pH is around 2 in the stomach. This low pH also **aids in the breakdown of connective tissue** and **helps kill many microorganisms (protection).**
    d. Pepsin: A digestive enzyme that is secreted as pepsinogen, pepsin is a proteolytic enzyme that is made in the precursor form because otherwise it would kill the cell that made it. Pepsin is activated by HCl. As a proteolytic enzyme, it initiates protein digestion. However, this does not break proteins all the way down to individual amino acids.
    e. Intrinsic factor: This molecule is essential for vitamin B12 absorption. Pernicious anemia: lack of vitamin B12 (needed for stroma of RBC) due to lack of intrinsic factor

iv. Regulation of secretions: Three phases.
    (1) Cephalic phase: Phase of gastric digestion in which the smell of something pleasant and/or thoughts of something tasty cause vagal impulses to be sent to the stomach to increase the motility and gastric juice secretion. This is an anticipatory response.
    (2) Gastric phase: Phase of gastric secretion in which mechanical stretching of gut causes an increased gastric motility, increased HCl secretion, and increased pepsinogen secretion. Also in this phase, the hormone **gastrin is released from stomach epithelial cells.** Gastrin is released into circulation, and then it comes back to the stomach to increase the secretion of HCl and pepsinogen. **The mechanism at work for the gastrin signaling is thought to be via histamine and histamine receptors:**
        a. H_2 receptors bind gastrin and subsequently signal the cell to secrete.
        b. For patients with ulcers, we block these H_2 receptors.
        c. Drugs (TAGAMET, ZANTAC, PEPCID): These are all used to block H_2 receptors and thus lower HCl secretion.
        d. Alcohol and caffeine seem to stimulate gastrin and HCl release.
    (3) Intestinal phase: This is the restraining phase. At this phase, we now begin working to keep from "gastric dumping" (moving contents into the duodenum before the duodenum is ready).
        a. Enterogastric reflex: This is regulated entirely in the nerve plexus (enteric nerve plexus). The food in the SI inhibits GI motility and secretions.
        b. Enterogastrones are substances that are released from the SI to inhibit stomach motility/secretion (somatostatin, glucagon-like peptide).
    (4) Nothing is really absorbed in the stomach except for a few rare things **(aspirin, alcohol).**
        a. Alcohol levels in the plasma are dependent on:
            - Amount consumed
            - Rate of absorption
            - Rate of conversion to acetaldehyde (usually in the liver, but also some in the stomach). The stomach has alcohol dehydrogenase (ADH) in higher concentrations than the gut, and thus if we can keep alcohol in the stomach longer, we can break more alcohol down before it reaches the blood. Thus, if we eat food with alcohol, the alcohol is less inebriating (particularly with fats, which cause the release of enterogastrones to slow GI motility). Also, note that women tend to be smaller and have a much less potent ADH (50% strength), and thus become inebriated much quicker and with lesser amounts.

V. Small Intestine: Called the "small" intestine because of its diameter, not because of its length. The SI is the longest and most convoluted of the intestines (about 20 feet).

a. **Anatomy**

   i. **Regions: There are three parts.**

      (1) **Duodenum:** Secretions from the pancreatic duct and the gallbladder (common bile duct) enter here.

      (2) **Jejunum**

      (3) **Ileum**

   ii. **Wall modifications in the duodenum:**

      (1) Within the wall are the **plicae circulares,** which are folds within the wall (large folds) that work to increase the surface area.

      (2) On the plica circulares are the **villi** (fingerlike projections), which project to the surface. There are different cells in the villi:

         a. **Mucus-producing cells**

         b. **Epithelial cells:** On the border of the epithelial cells are the **microvilli** (hairs on the peach—**"intestinal fuzz"**). Each epithelial cell may have up to 6,000 microvilli.

      (3) Within the microvilli are enzymes.

         a. **Enteropeptidase:** Activates trypsinogen to trypsin

         b. **Disaccharidases:** Break disaccharides down to monosaccharides

         c. **Aminopeptidases:** Break short chain amino acids down to individual amino acids

         d. **The enzymes within microvilli are responsible for the final digestion.**

      (4) Between adjacent villi are depressions (**Crypts of Lieberkühn, or intestinal glands**). These are very active areas. The epithelial cells within this region are "nurseries" for other cell types (they are the precursors to other cell types). We repair the surface of our SI every 3 days! Every minute, 100 million cells are formed here. These cells are especially vulnerable to **radiation/chemotherapy** (because these therapies target highly mitotic cells [cancer cells]). As a result, patients involved in chemotherapy or radiation therapy have resultant GI problems.

      (5) In the villi are **capillaries** (absorb amino acids and monosaccharides and transport them to the liver where that stuff may be metabolized) and **lacteals** (functionally, these are capillaries for the lymphatic system, which absorb chylomicrons).

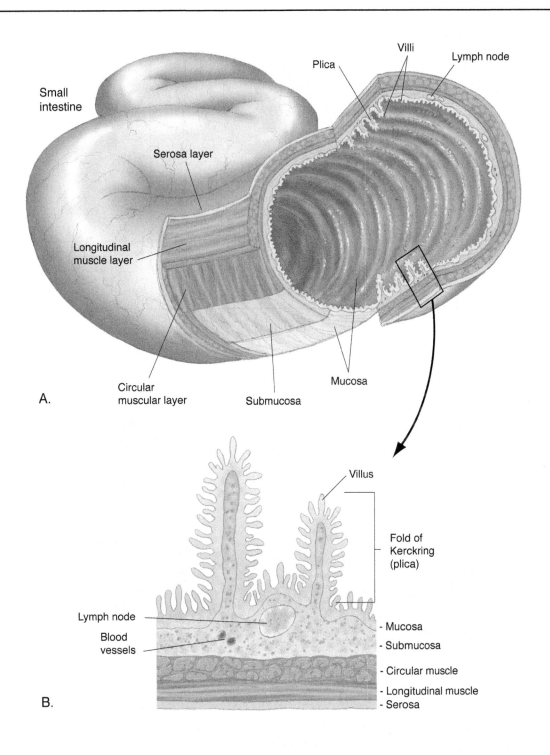

A.

Small
intestine

Serosa layer

Longitudinal
muscle layer

Circular
muscular layer

Submucosa

Mucosa

Plica

Villi

Lymph node

B.

Villus

Fold of
Kerckring
(plica)

Lymph node

Blood
vessels

- Mucosa
- Submucosa
- Circular muscle
- Longitudinal muscle
- Serosa

FIGURE 19.6

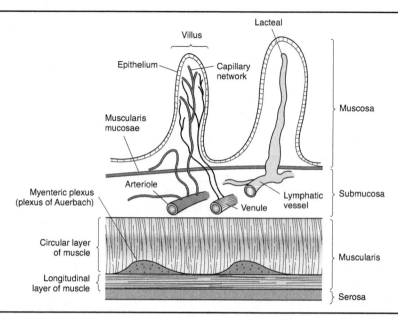

FIGURE 19.7

VI. Pancreatic and Liver Secretions

Figure 19.8 shows some of the relationships between the liver, gallbladder, pancreas, and small intestines: (1) When eating, parasympathetic stimulation increases bile production from the liver to the gallbladder. (2) Food/chyme moves from the stomach into the small intestines, where small intestine cells secrete hormones into the blood (e.g. CCK and secretin). (3) CCK travels via the blood to the gallbladder, where it stimulates the release of a bolus of bile into the cystic duct to aid in digestion of food in the small intestine (particularly fats - see Figure 19.9). (4) Secretin travels via the blood to the pancreas, where it stimulates the release of enzymes from the pancreas into the small intestine to aid in the digestion of food.

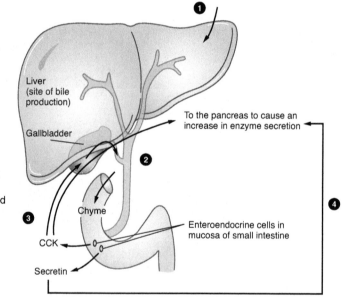

FIGURE 19.8

a. **Pancreas:** Exocrine and endocrine function. Secretions from the acinar cells find their way to the duodenum via the pancreatic duct. **Two exocrine secretions:**

i.   **Bicarbonate**—Part of a secretion that is a large volume, aqueous, and high in bicarbonate. This neutralizes acid chyme but has no digestive activity. The pancreatic enzymes have an optimal pH that is greater than 2.0.
ii. **Enzymes**—The pancreas produces enzymes in a small-volume, highly concentrated secretion that includes:
   (1) Proteases
      a. **Trypsinogen** is converted into trypsin by enteropeptidases (enzymes found in the microvilli).
      b. **Chymotrypsinogen** is converted into chymotrypsin via trypsin.
      c. **Procarboxypeptidase** is converted into carboxypeptidase by trypsin.
      d. Trypsin is responsible for the conversion into the active form of these enzymes.
      e. These proteases are responsible for the final digestion of proteins.
   (2) Lipases: Break fats into MAGs and FFAs (final digestion)
   (3) Amylases: Break down polysaccharides into disaccharides (*not* final digestion until reaching disaccharidases in microvilli)
   (4) Other enzymes (RNases, DNases, etc.)
iii. **Liver (bile salts):** Produce bile salts for the digestion of fats (see Chapter 11).
   (1) FFAs and MAGs are taken up into epithelial cell and then wrapped up in a protein coat to form a chylomicron. This is then taken up via the lacteal into the lymphatic system.
iv. **Regulation of Pancreatic and Liver Secretions:** Regulation by 2 main hormones:
   (1) Secretin
      a. Produced by epithelial cells of the SI
      b. Most predominant stimulus is the presence of acid chyme
      c. Less stimulated by bile and lipids
      d. This hormone goes to the pancreas to signal secretion of a bicarbonate-rich secretion. This is needed because the low pH of chyme is too far below the optimal pH of digestive enzymes from pancreas (especially lipases).
   (2) Cholecystokinin
      a. Produced by the intestinal epithelial cells of SI
      b. Most predominant stimulus is the presence of food (particularly fats) but also, to a lesser extent, vagal stimulation.
      c. This hormone causes the pancreas to release an enzyme-rich secretion.
      d. This also causes the gallbladder to release bile salts into the SI.

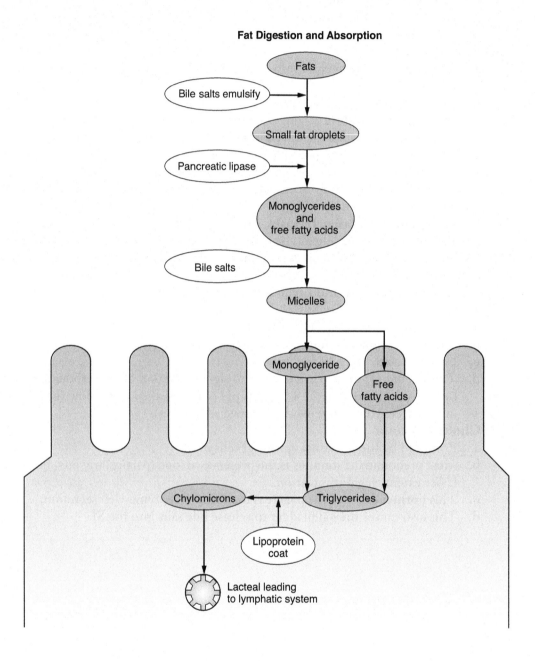

FIGURE 19.9

a.   **Fats**
   i.   Fats are reduced to monoglycerides and free fatty acids so that they are ready for absorption.
   ii. **Start with a large triglyceride:**
       (1) Bile salts emulsify fats.
       (2) Pancreatic lipases break fats down into monoglycerides and free fatty acids.

(3) Transported by the micelles to the epithelium of the SI, where they are absorbed.
(4) New triglycerides are synthesized in the epithelial cells.
(5) The triglycerides are wrapped in a protein coat to form chylomicrons.
(6) They are absorbed in the lacteal (part of the lymphatic system).

b. **Carbohydrates:** Pancreatic amylases break polysaccharides down into disaccharides, but they still are not ready for absorption. Disaccharides are broken down into monosaccharides via disaccharidases within the microvilli and then absorbed. They are then absorbed by the Na^+ co-transport system.

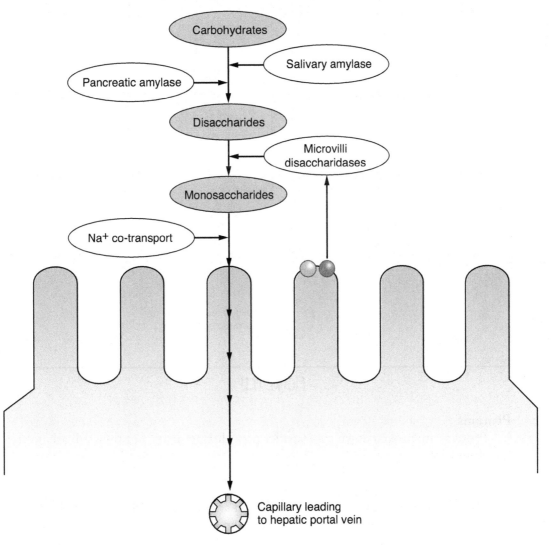

Carbohydrate Digestion and Absorption

FIGURE 19.10

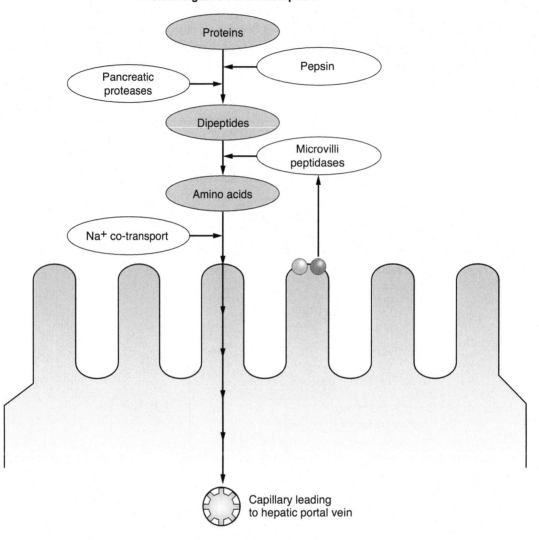

Protein Digestion and Absorption

FIGURE 19.11

c. **Proteins**
   i. Proteins are broken down into smaller peptide fragments, but they still are not ready for absorption.
      (1) Pancreatic proteases break down proteins into short-chain amino acids (along with pepsin in the stomach).
      (2) Amino peptidases break the short-chain amino acids into simple amino acids in the microvilli of the intestines. They are then absorbed via the Na^+ co-transport system.
   ii. They are absorbed into the capillaries, then the hepatic portal vein, and then into the sinusoids and the hepatic vein.

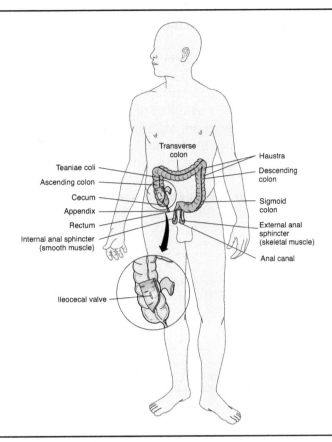

FIGURE 19.12

a. Where the small intestines join the large intestines is the **cecum** (pouch), from which the appendix hangs.

   i.   In humans, there is no known function for the cecum and appendix.

   ii. **Rabbits have a cecum: it does the same things as a cow's rumen.**

   iii. Most plants are protected by cellulose. Humans have no cellulases, but a cow's rumen and and a rabbit's cecum house microflora that do have this enzyme.

   iv. Rumen is anterior to the absorptive surface (there is microflora inside to break down the cellulose so that the animal can utilize the energy).

   v.   In rabbits, the same things happen in the cecum, but with one difference: the cecum is posterior to the absorptive surface. Thus, the rabbits are **coprophagic** (they eat their own feces to derive energy).

   vi. The appendix can become impacted with feces, causing a **fecalith (an infected appendix [appendicitis]).**

   (1) This can become infected and swell.

   (2) If it bursts, it will release fecal material into the peritoneal cavity, and peritonitis can result.

b. **LI: The ascending colon, transverse colon, and the descending colon**

   i.   Sigmoid colon at the end of the large intestines is followed by the anal canal and the anus.

 ii. The longitudinal muscle is reduced to three bands: **"taeniae coli"**
   (1) The taeniae coli remain partially contracted.
   (2) They pucker up into pockets called **haustra.**
   (3) The haustra help propel contents to the rectum.

c. Functions of LI
 i. There is absorption (mostly water and some Na^+).
 ii. Vitamin K is synthesized and absorbed here.
 iii. Two movements that move colon contents toward the rectum:
   (1) Haustral contractions—A slow shuffling of food
   (2) Gastrocolic reflex—More pronounced. After a meal there is a marked increase in contraction of the colon toward the rectum (gastro-colic reflex). The presence of food in the stomach causes the movement of food into the colon (integrated through the enteric nervous system).
   (3) Filling of the rectum then initiates an urge to defecate.

IX. Defecation

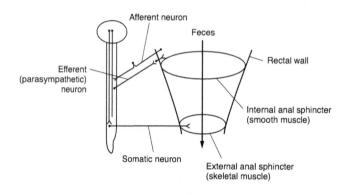

FIGURE 19.13

a. The defecation reflex:
 i. When the stretch receptors of the rectal wall are stretched, they send impulses to the rectum (via pseudounipolar neurons to the spinal cord and then efferent parasympathetic neurons to the rectum) that cause the internal sphincter to relax and the wall to contract.
 ii. The external sphincter is skeletal muscle and is innervated from the cerebrum (somatic efferent). We voluntarily control this muscle to defecate at appropriate times (contracting this also seems to eliminate the strong urge to defecate).

X. Pathologies

a. Ulcers: Typically the gastric mucosal barrier protects the surface of the stomach from the HCl and the digestive enzymes. This barrier is constantly formed by the mucosal layer and "destroyed" by pepsin. If the acid secretions and pepsin come in contact with the stomach wall, ulcers develop.
 i. **Gastric:** Mucosal barrier is affected
   (1) These are not as common as duodenal ulcers.
   (2) Bacteria indirectly produce ammonia, which breaks down the gastric mucosa, exposing the mucosal wall. (Treated with antibiotics)

     ii. **Duodenal:**
        (1) These are more common.
  b. **Zollinger-Ellison Syndrome:** A tumor usually in the pancreas. This results in abnormally large amounts of gastrin secretion. This causes secretion of HCl by the parietal cells of the stomach. The pH of the duodenum plummets too low. As a result, gastric lipases do not work very well. People with this do not digest fat very well, and they end up with excessive amounts of fat in their stool.
  c. **Inflammatory bowel disease:** Etiology is poorly understood. This is thought to be an auto-immune disorder. It presents with periods of remission and exacerbation.
     i. Crohn's disease: Inflammation of the SI (most commonly the ileum)
     ii. Ulcerative colitis: Inflammation of the colon
     iii. We can treat these symptomatically (with glucocorticoids to settle the immune system) or surgically (remove part of gut).
  d. **Appendicitis:** Due to impacted feces at appendix (fecalith).
  e. **Diverticulosis:** Outcroppings on the colon. This is more common in older populations (rare if <35 years of age). Patient could end up with diverticulitis (inflammation of the outcroppings). Usually, this can be treated with antibiotics. Patient may also want to avoid seeds, nuts, and other things that are hard to digest.

XI. Pharmacology
  a. Ulcer treatments
     i. **Neutralize the acid in the stomach:**
        (1) Aluminum hydroxide **(AMPHOJEL)**
        (2) MgOH plus AlOH **(MAALOX, MYLANTA)**
     ii. **H$_2$ Blockers (Histamine-2 blockers):** Decrease secretions of hydrochloric acid
        (1) (ZANTAC)
        (2) (PEPCID)
     iii. **Ulcer coater:** Puts coating on inside of stomach
        (1) (CARAFATE)
     iv. **Block hydrogen ion secretion**
        (1) (PRILOSEC)
        (2) (PREVACID)
        (3) (PROTONIX)
        (4) (NEXIUM)
  b. **Antiemetics:** These prevent nausea.
     i. Antihistamines
        (1) (ANTIVERT)
        (2) (DRAMAMINE): For motion sickness (best if taken prophylactically)
     ii. Anticholinergics
        (1) (TRANSDERM SCOP): For sea sickness; patch worn behind ear (best if taken prophy-lactically)
     iii. Central medullary: Work in the medulla. Used often with anticancer drugs (chemotherapy, etc.)
        (1) (TIGAN)
        (2) (ZOFRAN)
  c. **Emetics (Ipecac syrup):** Cause you to regurgitate. Given to patients who have ingested poison.
  d. **Antidiarrheals**
     i. **Local acting (KAOPECTATE, PEPTO-BISMOL):** Contain astringent that modifies the internal flora in some way. Their mechanism is poorly understood and they are **not endorsed by the FDA.**

ii. **Systemically acting** (not local acting)

    (1) Opiates (PARAGORIC): Contains opium; effective anti-diarrheals, but heavily regulated because they are very addictive

    (2) Synthetic opiates (LOMOTIL, IMODIUM): These are now available without a prescription.

e. **Laxatives**

    i. **Lactulose**—increases the osmotic pressure of the colonic contents.

SUGGESTED READINGS

1. Guyton, A.C., and Hall, J.E. 2011. Chapters 62–66 in *Textbook of Medical Physiology*, 12th ed. Baltimore: Saunders.
2. Koeppen, B.M. and Stanton, B.A. 2008. Chapters 26-31 in *Berne & Levy Physiology*, 6th ed. St. Louis: Mosby.

CHAPTER 19 QUESTIONS

1. _____ The lamina propria contains
   A. Longitudinal muscle
   B. Myenteric plexus
   C. GALT
   D. All of these
   E. Two of these

2. _____ A component of saliva
   A. Mucin
   B. Ptyalin
   C. Proteases
   D. All of these
   E. Two of these

3. _____ The oxyntic mucosa
   A. Contain chief cells
   B. Produce gastrin
   C. Produce intrinsic factor
   D. All of these
   E. Two of these

4. _____ Site of somatostatin production
   A. Pancreatic islets
   B. Hypothalamus
   C. Gastric pits of the PGA
   D. All of these
   E. Two of these

5. _____ The microvilli of the SI
   A. Produce lipases
   B. Produce enteropeptidase
   C. Produce disaccharidases
   D. All of these
   E. Two of these

6. _____ Cholecystokinin
   A. Is produced by intestinal mucosal cells
   B. Causes an enzyme-rich secretion by the pancreas
   C. Is stored in the gall bladder
   D. None of these
   E. Two of these

7. _____ An H-2 receptor blocker
   A. Zantac
   B. Prilosec
   C. Carafate
   D. All of these
   E. Two of these

For questions 8–10, choose one of the following:

        A. Increases or is greater than

        B. Decreases or is less than

        C. Has no effect or is equal to

8.   _____   The effect of secretin on pepsinogen secretion

9.   _____   The effect of glucagon-like peptide on gastric motility

10.   _____   The amount of gastrin secreted by a person suffering from Zollinger-Ellison syndrome, as compared to the amount secreted in a normal person

CHAPTER 20

Energy Metabolism

Energy: The capacity to do work (a system contains energy if it has the capacity to do work)

I. **Laws of Thermodynamics:**
   a. **First Law (Conservation of Energy):** Energy can neither be created nor destroyed, only transformed.
   b. **Second Law (Entropy):** Entropy of the universe is always increasing.
   c. **Third Law:** Absolute 0 can never be obtained.
      In lieu of the second law, can humans not decrease entropy by organizing their bodies? (That is, humans take food and form tissues and other things out of it—how is this formation of proteins and synthesis of tissues a decrease in entropy?)

   $$H + H \rightarrow H_2 \text{ (helium)}$$

   This releases a large amount of energy toward the earth in packets called photons.
   Plants with chlorophyll trap this energy.
   We, as humans, consume plants (directly or indirectly).

   Stoichiometry of glucose burning (1 to 1 ratio):

   $$C_6H_{12}O_6 + 6\ O_2 \rightarrow 6\ CO_2 + 6\ H_2O + \text{Energy}$$

II. **Manifestations of Energy:** The energy being released in a unit of time is called the metabolic rate. There are three manifestations of the metabolic rate.
   a. **Energy Storage:** Glycogen, fat, ATP
   b. **Work:** In mechanical movement/work—released energy
   c. **Heat**

III. Concept of Basal Metabolic Rate (BMR)

a. **Compare metabolic rates:** Metabolic rate will be much higher in someone exercising (at that time).

    i. **BMR:** The energy being released in a unit of time in specific conditions-at rest and post absorptive

    ii. Post absorptive we are storing very little energy.

    iii. While resting, the work rate approaches 0.

    iv. Heat production is the manifestation of energy release in an evaluation of BMR.

IV. Measurement of Metabolic Rate

a. **Direct Calorimetry:** Directly measures heat loss. Put a person in a chamber and measure the water jacket temperature change that surrounds the chamber.

b. **Indirect Calorimetry:** This is based on the stoichiometry of macronutrient usage.

$$C_6H_{12}O_6 + 6\ O_2 \rightarrow 6\ CO_2 + 6\ H_2O + \textbf{Energy}$$

    i. We can measure O_2 consumption by using indirect calorimetry and theoretically calculate how much heat (calories) they are producing.

V. Factors Affecting Metabolic Rate

a. **Size**

Animal	Mass (g)	O_2/hr
Shrew	4.8	35.5
Squirrel	96	98.8
Dog	11,700	3,870
Human	70,000	14,760
Elephant	3,833,000	268,000

Animal	Mass (g)	O_2/hr	O_2/g/hr
Shrew	4.8	35.5	7.4
Squirrel	96	98.8	1.03
Dog	11,700	3,870	0.33
Human	70,000	14,760	0.21
Elephant	3,833,000	268,000	0.07

    i. **Divide by body weight to get a better idea (lower chart).**

    ii. **On a per-gram basis,** a shrew's tissue takes more O_2 per gram. Thus, with the same mass of shrew tissue and elephant tissue—it would be *much* cheaper to feed the elephant tissue.

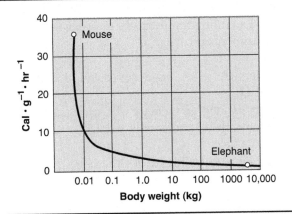

FIGURE 20.1

iii. **Endoderms:** Warm-blooded animals
iv. **Log function of metabolic rate as a log of body mass**

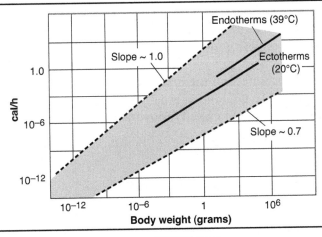

FIGURE 20.2

    (1) $VO_2 = a*M^b$ (equation of a straight line)
    (2) b = slope of the line
    (3) $M = ax^b$
    (4) $VO_2 = aM^{0.75}$ = slope of that line
  v. As the animal gets bigger, its metabolic rate gets bigger to the 0.75 power, *not* as a direct relationship (to the 1ˢᵗ power).
 vi. Why?
    (1) Geometry of a sphere
      a. A 10-g mass has more surface to volume than a 100-g mass.
      b. Yes, but this ratio changes to the 0.66 power, not to the 0.75 power.
    (2) Mouse vs. Steer metabolic rates
      a. Mouse with a comparably reduced metabolic rate of a steer: This mouse would need hair 20-cm thick to stay warm! There is probably some selective pressure against this.

  b. **Steer with a comparably increased metabolic rate of a mouse:** This steer would have skin that approached a boiling temperature in order for it to thermoregulate and keep a physiological core temperature.

  vii. LSD Story in Science

   (1) Scientists gave an elephant a 300-mg dose instead of an 80-mg dose!

b. **Temperature: Concept of Q_{10}**

 i. The Q_{10} effect: For every 10-degree increase (C), what is the resultant change in reaction rate? (In biological systems, the Q_{10} is usually ~2.2.)

c. **Age**

 i. As people grow older, their metabolic rate declines.

 ii. This will affect drug dosages for older people.

  (1) Cannot give an 85-year-old person the same drug dose when they were a 35-year-old.

d. **Hormones**

 i. Thyroxine has the most notable effect.

 ii. Thyroxine increases mitochondrial density and mitochondrial surface area.

e. **Food Stuffs**

 i. **Thermic Effect (specific dynamic action):** When different foodstuffs are processed in the body, there is an increase in the metabolic rate that is *not* the work of digestion.

 ii. Proteins have the greatest effect on metabolic rate (20% increase).

 iii. Carbohydrates have a good effect also (10% increase).

 iv. Fats have less than a 5% effect on metabolic rate.

f. **Activities**

 i. Activities have the greatest effect on the metabolic rate.

 ii. However, the greatest cost to the body is usually simply to maintain the BMR.

 How much energy have you expended (above your BMR) reading this textbook?

SUGGESTED READINGS

1. Guyton, A.C., and Hall, J.E. 2011. Chapter 72 in *Textbook of Medical Physiology*, 12th ed. Baltimore: Saunders.

INDEX

endometriosis, 262
endometrium, 249
endometrium cycle, 251
enteropeptidase, 300
ependymal cells, 44
ephedrine, 100
epicardium, 111
epididymis, 238, 239, 241, 243, 245
epidural space, 65
epiglottic cartilage, 182
epilepsy, 79
epimetrium, 249
epinephrine, 77, 100
epithelial cells, 294
erythroblasts, 196
erythrocytes, 195
erythropoiesis, 195–196
essential hypertension, 137
estrogen, effect on breasts, 261
estrogen, functions of, 251
estrogen, in endometrial cycle, 251
estrogen, in ovarian cycle, 250
estrogen, in parturition, 259
estrogen, in pregnancy, 257–258
estrous cycle, 248
ethanol, 160
excitability threshold, 18
excitatory post-synaptic potential, 49
exophthalmos, 280
expiratory reserve volume (ERV), 188
external nares, 181
extrinsic cascade system, 212
eyes, 93

F
fecalith, 307
ferritin, 198
fibrin, 213
fibrinogen, 213
filtration fraction, 150
first law of thermodynamics, 313
5-alpha-reductase, 240, 244
follicle-stimulating hormone (FSH), 223–225,
   246–247, 250, 252–253
follicles, thyroid, 276
follicular development, 249–250
follicular phase, 250
foramen of Luschka, 66
foramen of Magendie, 64, 66
foramen of Monroe, 66
forced expiratory volume (FEV)$_{1.0,}$ 188
forebrain, 55–59
functional residual capacity, 188
furosemide, 160

G
G-protein, 7
GABA, 77

gallstones, 207
ganglion, 45
gastric reflux, 297
gastrocolic reflex, 308
gastrointestinal tract, innervation, 95
genital fold, embryonic, 243
genital swelling, embryonic, 243
genital tubercles, embryonic, 243
germ cells, 240
gigantism, 233
glans penis, 241, 243
glaucoma, 93, 98
glial cells, 44
glioma, 78
glomerular filtration rate, 136, 153, 160
glucagon, 270
glutamate, 77
glycine, 77
glycolysis, 32
goiter, 280
Goldman equation, 16
gonadotrophs, 223
gonadotropin-releasing hormone (GnRH), 224, 246,
   254, 262
gout, 159
Graafian follicle. *see* mature follicle
granulocytes, 195
granulosa cells, 250, 252
Graves' disease, 280
growth hormone–inhibiting hormone (GHIH), 224
growth hormone–releasing hormone (GHRH), 224
guaifenesin, 199
Guillain-Barré syndrome (Landry's paralysis), 46
gut associated lymphatic tissue (GALT), 294
gynecomastia, 241, 260

H
Haldane effect, 194
Haustral contractions, 308
heart rate, 110
hemoglobin, 162, 191
hemophilia, 214
hemorrhoids, 132
hemosiderin, 198
Henderson-Hasselbach equation, 165
Henry's Law, 189
heparin, 213
hepatitis, 209–210
Herring Breuer reflex, 178
hindbrain, 60
hirsutism, 286
histamines, 77, 138, 189
hormone-sensitive lipase (HSL), 271
human chorionic gonadotropin (HCG), 244, 257
human chorionic somatomammotropin (HCS),
   258, 261
human growth hormone (hGH), 222, 224, 226,
   229–233

human placental lactogen (HPL). *see* human chorionic somatomammotropin (HCS)
hydrocephaly, 67
hydrochlorothiazide, 160
hymen, 249
hypercalcemia, 283
hyperplasia, 230
hypertension, 137
hyperthyroidism, 280
hypertrophy, 230
hypocalcemia, 283
hypospadia, 247
hypothalamus, 55, 59
hypothyroidism, 279–280
hypoxia, 193, 197

I

ibuprofen, 288
indirect calorimetry, 314
infant respiratory distress syndrome (IRDS), 185
inferior colliculi, 59
inflammatory bowel disease (IBD), 309
inguinal canal, 239
inhibin, 240, 246, 252, 253
inhibitory post-synaptic potential, 50
inspiratory reserve volume (IRV), 188
insulin, 155, 268–270
insulin-like growth factors (IGFs), 231
intercalated disks, 107
internal nares, 181
interstitial cells of Cajal, 295
intra-uterine device (IUD), 255
intrinsic cascade system, 212
intrinsic factor, 298
inulin, 155
iodide, 276–277
ipecac syrup, 309
islets of Langerhans, 268
isoproterenol, 101

J

jaundice, 199
juxtaglomerular apparatus (JGA), 136, 152

K

kidneys, innervation, 96

L

labia major, 243, 249
labia minor, 243, 249
lactation, 251, 258, 260–261
lactotrophs, 223
lactulose, 310
lamina propria, 294
LaPlace's law, 124–125
laryngopharynx, 182
laws of thermodynamics, 313

leukocytes, 195
levodopa, 58
Leydig cells, 240
limbic system, 60–61, 178
lipoproteins, chylomicrons, 208
lipoproteins, HLDL, 209
lipoproteins, LDL, 209
lipoproteins, VLDL, 208
lungs, innervation, 94
luteal phase, 250
luteinizing hormone (LH), 223, 225, 246–247, 250

M

macrocortin, 287
macula densa, 152
malignant hyperthermia, 37
mannitol, 160
mature follicle (Graafian follicle), 250
mechanoreceptors, 71
medulla, 55, 60
medulla ischemic receptors, 135
megakaryocytes, 195, 196
melanocyte-stimulating hormone (MSH), 223, 224, 226
melatonin, 227, 247
menarche, 248
meninges, 65
meningitis, 79
menstrual cycle, 248
mesangial cells, 153
metabolic acidosis, 162, 166–167
metabolic alkalosis, 167
methylxanthine, 160
micelles, 206
microglia, 44
microvilli, small intestines, 300
midbrain, 59
mifepristone (RU 486), 255
mitral valve prolapse, 124
mons pubis, 249
morning-after pill, 255
morula, 256
motor units, 24, 35
mucous neck cells, 298
Mullerian, 243, 244
Mullerian-inhibiting factor, 244, 249
multiple neurofibromatosis, 78
multiple sclerosis, 46
muscarinic receptors, 88
muscle fiber, 26–28
muscle pump, 132
muscle relaxants, 38
muscular dystrophy, 36
myasthenia gravis, 37, 99
myenteric nerve plexus, 295
myocardial infraction, 122–123
myocardial ischemia, 121
myocardium, 111
myofibrils, 26

progesterone, functions of, 252
progesterone, in ovarian cycle, 250–251
progesterone, in pregnancy, 257
progesterone, placental, 258
prolactin, 225, 261
prolactin-inhibiting hormone (PIH), 224
propranolol, 102
proprioception, 74
prostacyclins, 211
prostaglandins, 256, 259, 289
prostate, 238, 241, 243, 245
prothrombin, 212
pseudohermaphroditism, 286
pulmonary edema, 132
pulmonary fibrosis, 189
pulse pressure, 128
pyramidal pathway, 76

Q
Q_{10} concept, 316

R
Rathke's pouch, 222
reflex arc, 41
relaxin, 258
renal clearance, 147–148
renal failure, 159–160, 271
renal hypertension, 137
renin-angiotensin system, 136, 143, 153, 286
reserpine, 90, 100
residual volume (RV), 188
respiratory alkalosis, 166
respiratory unit, 183
Reye's syndrome, 79
rhythm method, 255
rickets, 283
RU 486. *see* mifepristone
rubrospinal pathway, 76
ryanodine receptors, 28, 31

S
salivary glands, innervation, 95
saltatory conduction, 46
sarcomere, 26, 27
sarcoplasmic reticulum (SR), 27, 108
satellite cells, 44
saxitoxin (STX), 19
schizophrenia, 61
Schultze, 268
Schwann cells, 44, 78
sciatica, 78
scrotum, 239, 243, 245
second law of thermodynamics, 313
secondary follicle, 250
secretin, 205, 267, 303
seizure disorder, 79
Selye, Hans, 287
seminal vesicle, 241, 243, 245

seminiferous tubules, 239, 245
septic shock, 138
serotonin, 77
sertoli cells, 240
shingles, 78
shock, 100
shock, circulatory, 137–138
shock, hypovolemic, 138
shock, neurogenic, 138
shock, vasogenic, 138
smooth muscle, 36
SNAP25, 25
somatostatin, 269, 270
somatotrophs, 222
spatial summation, 52
spermatic cord, 239
spermatogenesis, 239
spinal cord, 62–64
spinal nerves, 83
spinal nerves, classes, 67–68
spinal tap, 67
spinocerebellar pathway, 74
spinocerebellum, 60
spinothalamic system, 74
SRY region, 242
Starling's Law, 36
statins, 209
stercobilin, 199
steroidogenesis, 239, 240, 245, 246
steroids, 221
Stokes-Adams syndrome, 118
streptokinase, 214
stroke, 79
stroke volume (SV), 109, 114
strychnine, 77
subarachnoid space, 65
submucosal nerve plexus, 295
substance P, 77
succinylcholine, 37, 98
sulfonylureas, 272
superior colliculi, 59
supraoptic nucleus, 222
surfactant, 185
sympathetic nervous system, 83
synapses, 18, 48–52
synaptobrevin, 25
synaptotagmin, 25
syncope, micturition vasovagal, 138
syntaxin, 25
systole, 113

T
t-tubules, 27–28
T3. *see* triiodothyronine
T4. *see* tetraiodothyronine
tachycardia, 117
taeniae coli, 308
tegmentum, 60

CPSIA information can be obtained
at www.ICGtesting.com
Printed in the USA
LVOW02s1355301215

468185LV00003B/14/P

9 781465 209337